American Wilderness

American Wilderness

A New History

Edited by

MICHAEL LEWIS

OXFORD
UNIVERSITY PRESS
2007

OXFORD
UNIVERSITY PRESS

Oxford University Press, Inc., publishes works that further
Oxford University's objective of excellence
in research, scholarship, and education.

Oxford New York
Auckland Cape Town Dar es Salaam Hong Kong Karachi
Kuala Lumpur Madrid Melbourne Mexico City Nairobi
New Delhi Shanghai Taipei Toronto

With offices in
Argentina Austria Brazil Chile Czech Republic France Greece
Guatemala Hungary Italy Japan Poland Portugal Singapore
South Korea Switzerland Thailand Turkey Ukraine Vietnam

Published by Oxford University Press, Inc.
198 Madison Avenue, New York, New York 10016

www.oup.com

Library of Congress Cataloging-in-Publication Data
American wilderness : a new history / edited by Michael Lewis.
p. cm.
Includes bibliographical references and index.
ISBN 978-0-19-517415-1; 978-0-19-517414-4 (pbk.)
1. Human ecology—United States—History. 2. Geographical perception—United States.
3. Wilderness areas—United States—Public opinion. 4. Human beings—Effect of environment on—United States. 5. Public opinion—United States.
6. United States—Environmental conditions. I. Lewis, Michael L., 1971–
GF503A64 2007
304.2—dc22 2006048320

1 3 5 7 9 8 6 4 2

Printed in the United States of America
on acid-free paper

Contents

Contributors

Christopher Conte is an associate professor of history at Utah State University. He studies the historical transformations of land use and land cover in East Africa's highlands, as seen in *Highland Sanctuary: Environmental History in Tanzania's Usambara Mountains* (a Choice outstanding academic title for 2004).

Bradley P. Dean (1954–2006), former editor of the *Thoreau Society Bulletin*, was best known for compiling, analyzing, and editing Thoreau's unpublished manuscripts, including *Faith in a Seed* and *Wild Fruits*. He was working on Thoreau's unpublished "Indian Notebooks" at the time of his tragic and early death.

Mark Harvey, professor of history at North Dakota State University, is most recently the author of *Wilderness Forever: Howard Zahniser and the Path to the Wilderness Act.*

Kimberly A. Jarvis, author of *Nature and Identity in the Creation of Franconia Notch*, is assistant professor of history at Doane College.

Benjamin Johnson teaches in the History Department at Southern Methodist University. His publications include *Revolution in Texas: How a Forgotten Rebellion and Its Bloody Suppression Turned Mexicans into Americans*, and he is at work on a synthetic history of American conservation in the Progressive Era.

Michael Lewis, author of *Inventing Global Ecology: Tracking the Biodiversity Ideal in India, 1947–1997*, is an associate professor of history and the director of the Environmental Issues Program at Salisbury University.

Angela Miller, professor of art history at Washington University, includes among her publications *Empire of the Eye: Landscape Representation and American Cultural Politics, 1825–1875*, winner of the John Hope Franklin Prize (American Studies Association) and the Charles Eldredge Prize (Smithsonian Institution).

Char Miller, professor of history and director of urban studies at Trinity University, is the author and editor of numerous books, including the award-winning *Gifford Pinchot and the Making of Modern Environmentalism.*

Melanie Perreault, associate professor of history at Salisbury University, is the author of *Early English Encounters in Russia, West Africa and the Americas 1530–1614* and co-editor of *Colonial Chesapeake: New Perspectives.*

Mark Stoll, associate professor of history at Texas Tech University, is the editor of a book series on world environmental history and the author of two books, including *Protestantism, Capitalism, and Nature in America.*

Steven Stoll, associate professor of history at Yale University, is the author of *Larding the Lean Earth: Soil and Society in Nineteenth Century America* and *The Fruits of Natural Advantage: Making the Industrial Countryside in California.*

Paul Sutter, author of *Driven Wild: How the Fight against Automobiles Launched the Modern Wilderness Movement,* is an associate professor of history at the University of Georgia.

James Morton Turner is an assistant professor in the Environmental Studies Program at Wellesley College. His book *The Promise of Wilderness: A History of American Environmental Politics* will be published in 2008.

Donald Worster, Hall Distinguished Professor of History at the University of Kansas, is the award-winning author of numerous books and is currently working on a biography of John Muir.

American Wilderness

One

American Wilderness
An Introduction

Michael Lewis

Shining Rock. The Big Horn Mountains. Cumberland Island. Gates of the Arctic. Mojave. From the Absaroka-Beartooth to Yosemite, the names alone thrill. These are some of the 677 federally designated wilderness areas of the United States, the most carefully preserved landscapes in the nation. In many of these places, you can walk for days without seeing another person or any obvious signs of human artifice, should you choose to do so. Outside this federal national wilderness system, more wilderness is preserved in our national parks, our national forests, and the numerous other categories of federally managed land. Still more wilderness can be found in private property—from estates and hunting clubs to Nature Conservancy sites. In a world that is increasingly paved and groomed, such places are precious. And given the continued thirst of our consumerist society for resources, their existence is at first glance surprising. The nation that, arguably, most fully has embraced industrial capitalism and consumer culture, a nation whose wealth has been predicated on its ability to harvest and transform an unusually rich bounty of natural resources, simultaneously developed rationales and models for setting aside landscapes (often spectacular ones) as permanent wildlands. This book explores the apparently contradictory history of Americans and their wilderness.

Twenty-first-century Americans love wilderness. We idealize it, we romanticize it, we hike in it, we camp in it, we long to experience it. So many Americans want to enjoy wilderness that recreational specialists have devised "low-impact" camping techniques so that thousands of U.S. citizens can visit the same mountain or the same forest and feel as if they are the first to set foot in it. We name our automobiles after mountain ranges and rugged Western landscapes. We advertise beer with wilderness—"The taste of the Rockies," or "Come to the mountains, come to Busch beer." We hang pictures of wilderness on our walls. People dress every day as if they were heading out on a wilderness hike, carrying backpacks instead of briefcases, wearing polar fleece and hiking boots. Our national park system is the oldest in the world, and every year millions of Americans make pilgrimages to these spectacular, even sacred, sites. The U.S. environmental organizations that focus upon the preservation of wilderness and wild species have combined memberships reaching into the millions.

Of course, to say that "Americans love wilderness" is far too simple. At times, it seems that we might love our wilderness to death. Automotive gridlock exists not just in our cities; national parks such as Yellowstone, the Smoky Mountains, and the Grand Canyon can look like parking lots in the summer, with bumper-to-bumper traffic crawling along park roads and automotive exhaust clouding the sky. We show our love for our national parks by driving hundreds of miles to see them in RVs and SUVs that, at their best, travel fifteen miles per gallon of gas. Nowhere is our national schizophrenia more in evidence than in the ongoing debates over drilling for oil in the Arctic National Wildlife Refuge. Many Americans want to preserve the wilderness characteristics of this landscape, but they also drive the very cars—GMC Yukons and Toyota Tundras being the most ironically named—that make new sources of Arctic oil appear to be necessary.

When Americans are not on vacation, their lifestyles seem distinctly anti-wilderness. Suburban sprawl is eating up thousands of acres of land every year. National parks and wilderness areas are besieged by energy and mineral interests, by conflicts over wild carnivores leaving the parks, and by debates about forests and fires, to name just a few. Contemporary U.S. society is the most environmentally destructive in the world, if measured by resource use, energy consumption, per capita trash production, and other pollution measures. The United States, with approximately 4 percent of the world's population, emits roughly 25 percent of the global production of carbon dioxide. As we sit at home, the resources of the world's forests and fields surround us, from the wood of our furniture to the food in our refrigerators. Insofar as the countries of the developing world adopt a U.S.-style resource-intensive consumer culture, it bodes ill for the wilderness of the world.

So what does it mean to say that Americans love wilderness? Are we hypocrites? Our love for wilderness is tangibly visible on our national maps. The American national park system is perhaps our most globally accepted governmental idea, found (unlike democracy or the separation of powers) even in the cruelest dictatorships and single-party states. Yet national parks, with their tourist-oriented infrastructure of roads, visitor centers, and lodges, were not wild enough for many Americans, and in 1964 the Wilderness Act was passed, establishing national wilderness areas. According to this law, "wilderness, in contrast with those areas where man and his works dominate the landscape, is hereby recognized as an area where the earth and its community of life are untrammeled by man, where man himself is a visitor who does not remain." Wilderness areas have no roads, buildings, motorized vehicles, or campgrounds—not even chainsaws are allowed. They are held to a higher standard of nature protection than any other federal land in our country—and, at the time of the act's passage, in the world. Yet still, our national schizophrenia is everywhere in evidence, as the history of the implementation of both national parks and wilderness areas has been laced with setbacks and disappointments. Since the 1970s, many local communities have felt that a new wilderness area designated in their backyard would be a disaster. And love and legislation notwithstanding, the inexorable trend in the United States has been the steady net loss of wilderness and wild areas, even as a few new remnants are set aside.

Some historians have tried to explain these contradictions by analyzing wilderness politics—charting the growth of the wilderness movement over the years, the passage of legislation, and changing ideas about wilderness. Some others have attempted to explain why these political decisions were made and why American ideas about nature changed through time. They often look to other cultural trends and forces, and they try to study the American love of wilderness as a manifestation of larger cultural patterns. Still others have focused upon histories of the landscapes themselves, showing how particular natural areas have changed. They have all shared, however, the conviction that the key to understanding contemporary American interactions with wilderness is by studying the past and how those attitudes arose and changed through time.

The historicizing of love can be profoundly disconcerting to one *in* love. Wilderness advocates have often found the history of their movement, and their central life commitment, to be alienating, much as many religious people dislike religious history. This has been especially true when historians have attempted to analyze why people do what they do. It is one thing to chronicle *what* people do, and quite another to interpret *why* they do it. Who, after all, would want to be described as a person who buys wilderness art as part of a process of denial about her own environmentally destructive lifestyle, or as a person who supports wilderness bills for Alaska as a salve for his subconscious guilt over living in suburban sprawl? Not surprisingly, some wilderness advocates have come to think that wilderness history is irrelevant or even an impediment to their activism. As an example, when historians write about cultural constructions of wilderness—how wilderness is defined by the observer's culture—some activists are frustrated and reply that wilderness is a real thing, not an idea or a construction of culture.

Many an argument has begun because of a lack of understanding of a simple truth: wilderness is simultaneously a real thing and a human construction. Wilderness is difficult to define in part because it is a noun that "acts like an adjective," joining similar words, such as beauty or wisdom.[1] But unlike those words, wilderness refers to a completely nonsubjective, nonhuman, wild nature.

Although different people perceive wilderness in different ways, it is inaccurate to argue that wilderness (like beauty) is purely in the eye of the beholder. Historians are trained to study culture rather than forest ecosystems (more properly the province of scientists), and thus even environmental history often focuses more upon reconstructing past human perceptions and ideas than on reconstructing ecosystems. It is a mistake to read into that focus the assumption that human perceptions create reality.

For example, to a seventeenth-century Iroquois, a mid-Atlantic forest might look quite different than it would to a European explorer. The Iroquois might see the forest as a mosaic of habitats: some wilder wilderness sections, some periodically burned and managed for easy hunting of deer and turkey, and some former village or farming sites, now regenerating young shrubs and trees. The European explorer might see all of the forest as pure wilderness, a dark and mysterious landscape illustrating God's original paradise, a fearsome haunt of the devil, or even a collection of marketable commodities. Neither, probably, would be interested in joining a twenty-first-century backpacker voluntarily spending a week living in that same forest wilderness among the mosquitoes, snakes, and bears and cooking bad food over a fire, all in the name of relaxation. These three ways of seeing the wilderness are different, but the wilderness itself—the forest that they perceive and interpret—does not change.

It is also important to distinguish between wilderness and wildness. I have a tulip poplar tree in my backyard that is infested with aphids. Ants eat the highly nutritious excrement of aphids and in turn defend the aphids from marauding ladybugs that eat not the excrement but the aphids themselves—a life-and-death struggle over my back deck that is repeated every summer. This tree and insect community is wild; it is not wilderness. This is not because of the tameness of the tree (I did not plant it, and in fact it is too close to my house) or the lack of wild beasts (the insects are voracious and completely outside of my control). The tulip tree/insect community is not wilderness primarily because it is small. Wilderness is a concept devised by humans to define a particular type of wild environment—with its plants, animals, and ecosystems—and it is entirely appropriate to declare that wilderness, as distinct from wildness, must be large on a human scale. Wild nature can be found everywhere; wilderness cannot. The grass that grows in the seam of a concrete sidewalk is as wild as a bear in the Brooks Range of Alaska. No one, though, should question that one is in wilderness, the other is not, and learning to appreciate wildness need not replace an appreciation of wilderness. Too often, the two are treated as oppositional categories.

Just as wilderness has fascinated Americans over the last two hundred years—whether they responded with fear, loathing, respect, or love—every generation has had historians who wrote about nature or wilderness. Some historians have viewed the interaction between settlers and raw wilderness as the central reality of early American history. Francis Parkman, a historian at Harvard in the mid-1800s, wrote his most famous book about the Oregon Trail after walking a portion of the trail as research. Parkman saw the struggle of pioneers to settle the wilds of North America as the key story of his age, but he also mourned the end of the glorious, unsettled West (and was not at all flattering in his descriptions of the pioneers). Not coincidentally, Parkman taught at Harvard with such luminaries as Ralph Waldo Emerson, and he knew Thoreau as well.[2] At the turn of the twentieth century, there was another renowned generation of wilderness activists, including figures such as John Muir, Mary King Sherman, and Gifford Pinchot, and they also had their wilderness historian. In 1893, Wisconsin professor Frederick Jackson Turner delivered a paper, subsequently serialized in several books, entitled "The Frontier Thesis of American History." Turner argued that American civilization was in trouble, because the 1890 census had declared the frontier to be closed (the continental United States had achieved a minimal population density in all territories). He proposed that the frontier had acted as a release valve for the pressures of urbanization and as a forge for the national character. With no more empty wild lands, America seemed destined for a long, slow, decline. Turner's reading of the crucial role of wilderness (as frontier) in U.S. history helped to provide a justification for political action to save wilderness in the decades that saw the beginning of the national forest system, the expansion and formalization of the national park system, and the popularization of wilderness appreciation among the women and men of the United States.

In the 1960s, historians began to look more closely at the history of wilderness as an idea and the history of conservation efforts, as well as the earlier notion of wilderness as a forge for national character. Not coincidentally, the cultural context of that historical scholarship was the booming wilderness movement, highlighted by the 1964 passage of the national Wilderness Act. The incredible popularity of some of that historical scholarship—most particularly Roderick Nash's 1967 *Wilderness and the American Mind*—reflected not just the considerable quality of these writings, but also the degree to which Nash and his colleagues captured the environmental spirit of the age. That this scholarship is still popular forty years later reflects the continued relevance of the 1960s wilderness movement to the present. Nash's book, still in print and now in its fourth edition, is popular both in university courses and among general readers. Not only did Nash capture the wilderness pulse of America, he helped to define that pulse, so that subsequent generations of environmental scholars and activists discussed American wilderness in Nash's terms, with his examples, and with his heroes.

Nash argued that early Americans were predisposed by European culture and their Judeo-Christian heritage to see wilderness as evil, dangerous, and ungodly (in this, he was joined by the historian Lynn White, who wrote a celebrated 1967 article claiming that the Judeo-Christian world view of dominion, as expressed in Genesis, was incompatible with biocentric environmentalism).[3] As the forests of North America were cleared, however, and romanticism became the ascendant Western mode of thought, educated elites and urbanites began to attribute positive virtues to wilderness. Thus, wilderness appreciation originated among East Coast elites who were separated from wilderness by an industrializing civilization. By the twentieth century, this elite movement had spread into a "wilderness cult," a widespread and almost fanatical belief on the part of (still primarily middle-class and urban) Americans that wilderness would cure the problems of industrial society. This wilderness movement, however, was contested at its very core by preservationists, who saw wilderness as possessing intrinsic worth—Nash's heroes, Thoreau, Muir, and Aldo Leopold—and by conservationists, represented by Gifford Pinchot and his followers, who saw the conservation of wilderness as only a strategy for prolonging access to key natural resources for human use. In the twentieth century, then, Nash's history of wilderness was primarily the history of the conflict between these two groups. To Nash and his followers, the contradictory American behaviors listed at the start of this chapter represent a society in transition, where true wilderness lovers are still a minority in a population that is gradually moving away from more anthropocentric (human-centered) ways of seeing nature. Thus, different Americans are spread over a spectrum of wilderness appreciation, from high to low, with many still mixed up in the middle (and thus conflicted in their actions).

Insofar as Nash's ideas sound familiar, that reflects the degree to which they, and the cultural attitudes they reflect, continue to dominate popular American conceptions of wilderness. Historian Samuel P. Hayes makes an interesting contrast to Nash. Like Nash, he considered the beginnings of the wilderness conservation movement in 1890–1920 in *Conservation and the Gospel of Efficiency* (1959). Unlike Nash, Hayes viewed the conservation movement debates and changes firmly within the context of the Progressive Era, arguing that wilderness activism was not motivated solely by appreciation for the wilderness, but often by progressive concerns with order, rationality, and permanence. While historians have found Hayes's insights useful, he never gained as much of a following among the wilderness activists of the 1960s, who read Nash's work in such large numbers. His thesis did not dovetail as neatly with the goals of the wilderness movement, and perhaps activists found his larger historical concerns less relevant to their activism.[4]

In the decades since Nash's path-breaking book, dozens of historians followed in his footsteps and studied different aspects of the history of American wilderness ideas and politics. Practically every chapter of his book became the subject of numerous dissertations and monographs, and wilderness history became a key component of the growing discipline of environmental history. As is inevitably the case, as more scholars studied this history, it became increasingly complex. Among the gradual modifications of Nash's original argument, they suggested that the Christian legacy was not only subduing the earth but also saving God's creation; farms were sometimes as destructive of wilderness as industry; women played a key role in conservation efforts; not all wilderness activism began with urban elites; American Indians inhabited most of the "virgin wilderness"; not all European settlements shared the same values as New England; the conservation-preservation dichotomy was too stark, rigid, and simplistic; "wilderness" itself was not necessarily best studied in the history of wilderness areas. By the 1990s, a growing number of historians argued that these findings called for a different interpretation of wilderness and its history. Among these wilderness revisionists, none has been more influential than William Cronon, particularly through his 1995 essay, "The Trouble with Wilderness; or, Getting Back to the Wrong Nature."[5]

Cronon's essay argued that the traditional definition of wilderness (a pristine landscape untouched by human hands) was part of a persistent Western dualism juxtaposing wilderness and civilization, nature and human, as pure opposites. Ultimately, Cronon speculated, this wilderness idea actually worked against the emergence of an environmental movement that could make industrial societies livable. By claiming that any human touch (or, less extreme, any sign of industrial civilization) degrades a landscape beyond being worth saving, there is no need to focus upon improving the local environments in which we all work (and whose degradation is so economically attractive). We can save distant pure landscapes and abuse those on which we live. This allows many Americans to be good wilderness-loving environmentalists while they continue to participate in environmentally destructive development at home. In this view, wilderness preservation is not a reaction against industrialization, but actually enables the process. Wilderness is the cultural sleight of hand that makes our (often grotesque) contemporary development and exploitation mentally acceptable. He concluded that Americans need to spend more time focused on wildness—the wildness that can be found in the middle landscape that blends the human and the natural—than on a pure wilderness that exists outside of history.

Cronon's essay (and the other essays in the book of which it was a part) helped to ignite a raging debate that quickly moved from the pages of academic journals to newspapers and popular periodicals. Was wilderness a problematic concept? Was it the "wrong nature," as Cronon's title implied? Cronon had acknowledged that his essay was, in places, speculative, and some of his ideas were so sweeping as to be ultimately impossible to prove or disprove. Unfortunately, many of the arguments resulting from the essay eschewed subtlety (as was true of the essay's title as well, though the essay itself was far more nuanced). Not all responses were polemical, however, and at its best the controversy encouraged a number of people to reconsider the relationship between wilderness and American culture. The best arguments and essays (by historians, philosophers, and environmentalists) were collected in *The Great New Wilderness Debate* (1998).[6] As that volume demonstrates, both the traditional and revisionist interpretations are useful in understanding the past; any attempt to analyze the complexity of changing U.S. culture inevitably comes to realize that no one explanation is sufficient for any historical phenomenon. There is not just one monolithic American mind or way of seeing—or of loving—wilderness.

The new wilderness historians (those writing since 1990 and particularly those who began their work in the years immediately before and after the wilderness debates) share several characteristics. They are more attuned to power disparities and the politics of race, class, and gender than their 1960s predecessors. They are often concerned with U.S. overconsumption, the dismal state of our cities and urban sprawl, and America's role as the world's top polluter. And they are almost always people who have grown up enjoying wilderness in the United States—hiking and camping in the landscapes preserved by earlier wilderness activism. Their work has, however, been received by many contemporary wilderness activists as antithetical to the goal of preserving more wilderness on earth. Dave Foreman, a leading wilderness advocate, playfully suggested that Cronon and other wilderness historians had come up with their ideas "as they hold hands in a darkened room around a séance table, trying to hear voices from the misty shades of Jonathan Edwards and Henry David Thoreau." They were fixated on the past rather than the crises of the present. More seriously, he continued, "I have spent my life fighting the lies, blather, and myths of extractive industry about wilderness. I have concluded that their pitiful arguments against wilderness are actually more legitimate, rational and grounded in reality than those of the postmodern deconstructionists" (he means the historians).[7] Not surprisingly, then, most of the new wilderness histories have not received anything like the popular response afforded to Nash's *Wilderness and the American Mind*. Quite simply, the new wilderness historians are not telling stories that many environmentalists want to hear.

The new wilderness historians tend to argue that their work reflects the natural maturation of any academic subfield, particularly a subfield born of the concerns of a social movement. Over time, wilderness historians have deepened their analysis, discovered new sources and stories, and inevitably acquired a critical distance from the movement not initially possible in the first heady days of activist-fired academic inquiry in the 1960s. Nonetheless, it is still possible to see a cultural context for the new wilderness historians that is revealing. Just as Nash and his cohort wrote wilderness histories that bore the imprint of 1960s environmentalism, contemporary wilderness historians reflect a series of concerns and events that have emerged since the 1980s.

Rather than reflecting the exuberance of the passage of the Wilderness Act in 1964 and the ascendant environmental movement, the new wilderness histories arose in a different set of contexts. In 1987, the United Church of Christ's Commission for Racial Justice issued a report, "Toxic Waste and Race in the United States," that helped to bring national attention to the environmental justice movement. Environmental justice activists critiqued the ways in which traditional environmentalism had ignored the plight of the less privileged. What did wilderness conservation have to do with the poor or with racial minorities? they asked. On the opposite side of the political spectrum, the 1980s saw the Sagebrush rebellion, a conservative backlash in the Western United States against federal land management policies that were believed to favor elite Eastern interests and to ignore local concerns.

Internationally, there was a postcolonial backlash against what many developing-world activists, scholars, and government leaders referred to as the inappropriate imposition of U.S.-style wilderness preservation in the third world. These complaints were most famously voiced by Ramachandra Guha in his 1989 article, "Radical American Environmentalism and Wilderness Preservation: A Third World Critique," which argued that U.S. wilderness ideas were not appropriate in Asia and Africa.[8] At the UN-sponsored Rio Conference in 1992, representatives from the developing world roundly criticized what they saw as first world hypocrisy in proposing wilderness preservation in the developing world while not curbing first world resource consumption. In sum, historians writing about wilderness in the 1990s could not help but be aware of the voluminous critiques—at home and abroad—of the legacy of wilderness ideas. Whether they agreed or not, inevitably the debate influenced the way they conceived of and presented their scholarship.

By this point, the cumulative work on U.S. wilderness history fills several library shelves. The time has come to bring this new scholarship together and reframe the history of U.S. attitudes and actions toward wilderness in light of the last forty years of historical inquiry and to move past the wilderness debate of the last decade. The fourteen contributors to this volume do not agree on all points, but they do agree on the importance of drawing on the best ideas from both sides of the wilderness debate. Many of the traditional wilderness heroes and events are found here—Thoreau, Muir, Leopold, Hetch Hetchy, Echo Park, and the Wilderness Act—with new historical evidence and a wider historical context that makes them seem new. Conversely, there are several people, ideas, and events that will not be as familiar, from the way English settlers feared that wilderness might affect their bodies, to the agricultural roots of conservation, to the hiking women of the Appalachian Mountain Club, among many others.

There are five primary threads drawing these essays together. The most basic is a history of the growing wilderness movement, looking both at differing visions of what conservation might entail and at the movement's politics itself. Intertwined with that story, though, is the countercurrent of how the preservation of wilderness has negatively affected peoples as diverse as backwoods settlers, American Indians, and Maasai pastoralists. Analyses of the changing ideas that both shape and reflect American wilderness thinking—from religion, philosophy, agrarian conservation, art, literature, and science—constitute a third theme for the volume. A fourth key theme is how wilderness is linked to the nation, and nationalism, both in the United States and overseas. And woven through all of these stories is the backdrop of the expansion of U.S. agriculture, industry, population, and settlement over the last four hundred years and the resulting environmental degradation. As several chapters demonstrate, the ecological transformations that occurred in the American wilderness led to cultural and political responses.

The wilderness ideas and practices of the United States have been widely imitated around the globe and, in many cases, with striking historical parallels. Has this wilderness idea traveled so broadly because it fit into similar historical, cultural, and ecological contexts throughout the world, particularly, the emergence of nation-states and nationalism concurrent with industrialization and environmental degradation? Or, perhaps, was it linked to the spread of liberal democracy, as Donald Worster argues in this volume's epilogue? Though crucial to U.S. identity and history, the wilderness idea has never been just an American idea. Rather, it was derived from the shared human experience of modernity—the initially Euro-American, then global, experiences of the scientific revolution, exploration, colonialism, industrialization, and the dramatic transformation of the natural world. From Parkman to the present, U.S. historians have been convinced that to understand American wilderness is to understand a crucial part of America. Perhaps they have undersold the importance of wilderness history; perhaps to understand wilderness is to understand part of the more global history of modernity and its discontents: our values, our hopes, our blind spots, and our fears, overlaid on a rapidly changing planet.

Notes

1. Roderick Nash, *Wilderness and the American Mind*, 4th ed. (New Haven, Conn.: Yale University Press, 2001), 1.

2. Francis Parkman, *The Oregon Trail: Sketches of Prairie and Rocky Mountain Life* (New York: Oxford University Press, 1944).

3. Lynn White, "The Historical Roots of Our Ecological Crisis," *Science* (March 7, 1967): 1203–7.

4. Samuel P. Hayes, *Conservation and the Gospel of Efficiency: The Progressive Conservation Movement, 1890–1920* (Cambridge, Mass.: Harvard University Press, 1959).

5. William Cronon, "The Trouble with Wilderness; or, Getting Back to the Wrong Nature," in *Uncommon Ground: Toward Reinventing Nature*, ed. William Cronon (New York: Norton, 1995), 69–90. The other contributors to that volume make similar arguments, as do popular books like Michael Pollan, *Second Nature: A Gardener's Education* (New York: Dell, 1991).

6. J. Baird Callicott and Michael Nelson, eds., *The Great New Wilderness Debate* (Athens: University of Georgia Press, 1998).

7. Dave Foreman, "The Real Wilderness Idea," in *Wilderness Science in a Time of Change Conference*, vol. 1, *Changing Perspectives and Future Directions*, ed. David Cole, Stephen McCool, Wayne Freimund, and Jennifer O'Loughlin (Ogden, Utah: U.S. Department of Agriculture, Forest Service, Rocky Mountain Research Station, 2000), 32–38.

8. Ramachandra Guha, "Radical American Environmentalism and Wilderness Preservation: A Third World Critique," *Environmental Ethics* 11 (1989): 71–83.

Two

American Wilderness and First Contact

Melanie Perreault

The gods delivered a message ten years before the destruction began. A bright light pierced the dark of midnight, awakening the Nahua villagers. Frightened men and women looked to the east, where they saw a vision that "was like a flaming ear of corn, or a fiery signal, or the blaze of daybreak; it seemed to bleed fire, drop by drop, like a wound in the sky." Other bad omens followed: lightning struck temples, lakes suddenly boiled, and fishermen saw strange animals in the waters surrounding Tenochtitlan. But despite appeals to the most knowledgeable Aztec seers and magicians, no one could penetrate the meaning of the signs etched in the wilderness. The people went back to their cornfields, palaces, and marketplaces, but a general sense of unease settled on the land. Only later would they see that the disturbances were portents of the cultural and environmental devastation that would be set in motion upon the arrival of men whose sole interest in the land seemed to be in gathering gold.[1]

Thousands of miles to the northeast, and just over a hundred years later, another group of men and women looked warily at the landscape rolling out in front of them. The dark woods towered ominously over the coastline, harboring wild beasts and, from the few glimpses they had seen, wild men. As God's chosen people, they could take comfort in his protection—after all, they were there to establish a godly example for the rest of the world. But with such a blessing came a burden: the Devil was determined to prevent their success. As they scrutinized their new land for the first time, they could not help but wonder: did the Devil lurk in the woods, waiting to thwart their efforts? Would living in such a savage environment release the inner wilderness that all men and women restrained only through constant struggle?

Just to the south, a radical transformation had taken place in the previous twenty-five years. To be certain, the land had experienced earlier changes due to global environmental fluctuations and human intervention. The arrival of humans more than ten thousand years earlier was a significant turning point, made even more dramatic when they began domesticating crops. But nothing compared to the changes wrought in the two decades after a group of English settlers began building a fort on an island they called Jamestown. Here, each plant, fish, and tree was converted mentally into gold and silver. Entire fields that had once provided sustenance for human residents were now the home of a noxious weed grown on a scale never before seen in America.

Whether it was an Aztec premonition from 1511, a Puritan choking down his anxiety as he approached Massachusetts Bay in 1630, or a Virginian calculating next year's tobacco market in 1632, all of these confrontations with the American wilderness reveal the centrality of nature during the first meetings of Europeans and Native Americans during the early contact period. Europeans depicted American wilderness as a virtual paradise, a commodity-producing warehouse, a frightening malevolent entity, or a blank slate waiting to be brought to its full potential. But with very few exceptions, Europeans did not encounter a raw, untamed wilderness in America; they naturally established colonies in environments most fit for human occupation, where Native Americans already lived. And where Native Americans went, they altered the wilderness and transformed it into something else. As an *idea*, the notion of an untouched wilderness held great significance for Europeans and Native Americans alike during the early colonial efforts, but as a literal place, it did not exist, at least not in the areas where sustained contact took place. Certainly, some areas of North America were considerably more wild than others, however, and the forests and unfenced fields of the new land stood in stark contrast to the landscape of overdeveloped Western Europe.

While recognizing, then, that the American landscape was a dynamic entity shaped and reshaped by the forces of nature and humans over thousands of years, we must acknowledge that there were many wildernesses at the point of first contact between Europeans and Native Americans. This chapter examines the role of wilderness as physical space and ideological construct in the early contact period, focusing on case studies of the three stages of the contact experience. The first stage was defined by the Spanish efforts in the circum-Caribbean, when conquistadors expressed astonishment at what they saw in descriptions of the American environment heavily influenced by their desire for exotic goods, especially gold. During the second stage, other European powers sought to get a piece of the Spanish wealth and sent reconnaissance voyages up and down the eastern coast of America. The men involved in these voyages assessed the wilderness as a commodity, moving beyond the early Spanish obsession with gold to identify other material resources. In the third stage, Europeans established permanent colonies and confronted the wilderness on a new level. Here, the ideological and physical wilderness merged most clearly in a fear that prolonged exposure to the

American environment might cause Europeans to degenerate by incorporating the surrounding wilderness into their bodies and souls. In each case, Native Americans had their own notions of wilderness and their place in it, confounding European efforts to reduce them to a passive, essentially subhuman role.

The Spanish Encounter a "Virgin" Wilderness

One of the most enduring myths of American history is the European discovery of a virgin wilderness—an unpeopled land of pure nature. Of course, there were people in the Americas, but in a subsidiary to this myth, Americans have idealized the "noble savage," the Native American who walks gently through the woods of an untouched wilderness, careful not to alter the natural environment in any way lest he or she disturb nature. In deemphasizing the impact of Native Americans on their environment, the noble savage myth ultimately denied their humanity.[2]

Supporters of the noble savage myth have drawn a sharp distinction between the animistic world view of Native Americans and a detached materialistic European perspective that viewed nature as either a commodity or an obstacle. Yet, while there were important differences, at the time of first contact the gap between the two was not as wide as we might think. Men and women on both sides of the Atlantic followed carefully crafted rituals to appease natural spirits and ensure a good harvest. And extensive trade networks and markets to sell surplus and exotic goods existed in Europe and in the Americas long before 1492. The fifteenth and sixteenth centuries were a transitional period in the European intellectual understanding of nature, a bridge between medieval conceptions of the world and an Enlightenment movement that attempted to explain nature with scientific objectivity. The European voyages of exploration and the sudden revelation that there were plants, animals, and peoples whose existence had only been imagined played no small role in encouraging new thinking about wilderness.

During this transitional period, the separation of Christianity, science, and traditional pagan beliefs was not as complete as it would become later. Scientists often confidently asserted beliefs that we would categorize as magical today. European universities hosted debates about the efficacy of magic well into the seventeenth century, and reports of dragons and unicorns were treated seriously. English peasants danced around maypoles to energize the earth and lit bonfires to encourage fertility. Almanacs recorded stories of the birth of a horned child, and the appearance of a comet was cause for great reflection as to its meaning. In popular folklore, untamed forests were the home of dangerous animals, disorderly plants, and, most disturbing, a figure known as the "wild man." A frequent character in literature, paintings, and children's stories, the wild man was "a symbol of incivility, of near bestiality, of untamed nature." When Europeans set sail across the Atlantic in the fifteenth century, they brought with them a preconceived notion of what wilderness was and what kinds of people would live there.[3]

The earliest European accounts of the American environment seemed to confirm their belief that the land was an untouched wilderness inhabited by noble savages. There was little consensus, however, about what "wilderness" meant. The men who joined Hernando Cortes's expedition to the Valley of Mexico in 1519 made no pretenses about their interest in America—they were there to gather as much gold as they could find. Bernal Diaz del Castillo, a soldier who accompanied Cortes, described the scene as the Spanish got their first glimpse of the Aztec capital, Tenochtitlan: "we saw so many cities and villages built in the water and other great towns on dry land. . . . we were amazed and . . . some of our soldiers asked whether the things that we saw were not a dream." So many unexpected sights confronted the Spanish visitors—beautiful floating gardens, huge temples, and a dizzying array of people and animals—that they hardly had words to describe the scene. The vision, of course, was not a dream but was the end product of hundreds of years of human alteration of the landscape.[4]

Far from being an untamed wilderness, the Valley of Mexico was the site of one of the most ambitious wetland cultivation projects in human history. Though the natives who built the great cities lacked modern technology, they managed to transform the region using a dike and sluice system to construct the *chinampas* (artificial planting beds) that came to characterize Aztec agriculture. The Indians began constructing the *chinampas* in the first millennium C.E., gradually expanding until a population boom in the mid-fifteenth century forced a rapid increase in cultivation. What appeared to the Spanish to be floating gardens were actually rectangular plots created by alternating layers of mud and plants along the swampy edges of lakes, then securing them with the roots of willow trees. These gardens eventually covered more than 30,000 acres in the areas surrounding Tenochtitlan alone. This intensive and extensive agriculture was necessary to supply food for the unproductive classes (such as soldiers and politicians) who were essential to the centralized state system that allowed the Aztecs to dominate the region.[5]

Such a massive transformation did not come without environmental and human costs. Soil erosion and sedimentation were significant problems in Mexico and the Mayan lowlands long before the first Europeans arrived. Yet there is little question that the modification and degradation of the environment increased significantly under Spanish rule as immediate profits became more important than sustainability. Domestic animals accompanied Spanish expeditions into regions that were not suitable for large grass-eating quadrupeds, but ranchers simply moved their herds onto fresh lands after they had exhausted the local grasses. Spaniards in search of silver mines and ranching lands pressed ever northward, hoping for a source of wealth in the wilderness.[6]

Spanish explorers pushing into the American Southwest in the late sixteenth century did not find the huge urban centers of the Valley of Mexico, but even this semi-arid environment displayed the marks of human occupation. Here, the Hohokam and other Native Americans transformed the wilderness to support a large population in a marginal environment. From the third to the fifteenth centuries, the Hohokam built huge irrigation networks, even larger than those of the Aztecs. These networks allowed a substantial community to develop in what is today southern Arizona by supplying the water necessary to grow food. The sustainability of the system over a long period is questionable, however. The Hohokam disappeared in the mid-fifteenth century, perhaps due to the inability of the environment to support a large human population. The evidence of a massive depopulation connected to an environmental disaster is scanty, but there are intriguing indications of such an event. The word *Hohokam* is a Pima Indian word meaning "all used up," perhaps referring to water depletion and the subsequent drop in population.[7]

While the massive engineering projects of the Aztec and Hohokam peoples are dramatic examples of Native American transformation of the wilderness and are the easiest to find in the archaeological record, other peoples adopted different strategies to alter the environment around them to make it suitable or more comfortable for human habitation. Cultures that relied primarily on hunting and gathering for their subsistence did not build large cities or develop extensive markets; they never developed the surplus that a densely populated, centralized society required. Still, it would be a mistake to view these peoples as passive recipients of whatever nature happened to supply them with to survive. Instead, many of them used one of the most powerful and most pervasive tools of environmental transformation: fire.

Lightning strikes in dry forests are part of the natural progression of many ecosystems; intentionally set fires, however, reflect human intervention and manipulation of nature to various ends. Both agricultural and hunting and gathering cultures utilized controlled fires to alter the environment for their benefit. Fire returned valuable nutrients to the soil, cleared underbrush to facilitate movement, and removed vermin and disease from the surrounding area. Fired lands also encouraged the growth of edible plants, such as strawberries and blackberries. Hunters used fire to drive game animals into a confined space where they could be easily harvested. Indigenous burning practices were so widespread and so significantly shaped the American landscape that even in regions that were sparsely populated, humans were active agents of environmental change. European accounts of Indians from California to Roanoke Island noted that the Indians regularly burned forests and fields.[8]

Native Americans on the eastern coast of North America practiced a wide variety of subsistence activities that affected the environment to varying degrees. An agricultural revolution took place beginning around 1000 A.D., when a period of steady warming allowed the development of the "three sisters" (corn, beans, and squash) that dominated east coast agriculture at the time of European contact. In order to grow the three sisters, the Indians had to first perform the labor-intensive task of clearing the fields. In lightly wooded areas, the process was relatively simple—a controlled burn would remove much of the overgrowth. Wooded areas presented a greater challenge, particularly given the stone tools at their disposal. A combination of cutting and burning the trees over the course of a year or two could clear a field well enough to allow planting.[9]

Reconnaissance Voyages and a Wilderness of Commodities

When European colonists arrived in eastern North America in the late sixteenth and early seventeenth centuries, they encountered an environment already significantly transformed by the natives who lived there. Yet in their letters back home and in their promotional pamphlets, European authors described a land stuck in stasis, waiting for "civilized" peoples to come and develop it into its full potential. Some authors viewed the environment as a "howling wilderness," a desolate and dangerous setting in which to test their religious convictions. Others emphasized the economic potential of the land, calculating the financial possibilities of each tree, fish, or mineral. French fur traders competed with Jesuit missionaries for the attention of Indians in the Great Lakes region, while English and Dutch settlers struggled to survive in unfamiliar settings farther south. African laborers were forced to migrate to North America, lending their agricultural expertise to an inhuman system that valued profits above all else. Given the human and ecological diversity of the region, it is not surprising that there was no single view of the American wilderness during the early colonial period. Europeans were unified in believing that America *was* a wilderness, but what to do with it was another matter. Christianity, paganism, emerging capitalism, and early natural science all played a role in European efforts to explain (and thereby control) the perceived wilderness around them. Whatever their motivations, the colonists transformed, and to a significant degree were themselves transformed by, the eastern third of North America in the late sixteenth and early seventeenth centuries.

During this great age of exploration of the Atlantic, Europeans came into contact with numerous peoples and cultures, and they had to explain how such diversity could exist. As Christians, they agreed that all humans had to have originated from Adam and Eve, but what could explain the undeniably different skin tones and physical appearance of various Europeans, Africans, and Indians? In an age before scientific racism offered fixed hierarchies of race, European scholars turned to the most logical explanation that would not challenge biblical authority: while all humans lived together in the remote past, as time progressed they moved away from each other and began a process of differentiation. Removed from their environment of origin, people quickly began to degenerate and take on the characteristics of their new lands. People living in a wilderness, then, would be wilder than those living in domesticated environments. Natural philosophers argued that climate was the primary culprit in modifying people's bodies, but other environmental factors, such as the water people drank, could impart regional differences. The physical and cultural differences between peoples were a result of environmental influences rather than the much more dangerous suggestion of polygenesis.[10]

National identity, indeed the very constitution of their bodies, rested largely on the natural world around them, and Europeans scrambled to demonstrate that their homelands were the best on the globe. English writers were quick to emphasize that while other people had to maintain constant vigilance to avoid an encroaching wilderness, their environment was relatively tame and benevolent, well suited for hosting a superior civilization. Indeed, Englishman William Harrison argued, "[I]t is none of the least blessings wherewith God hath endued this island that it is void of noisome beasts, as lions, bears, leopards, wolves, and suchlike." While the unfortunate residents of untamed lands had to devote considerable attention to fending off the attacks of violent beasts, the English could rest assured that no such animals inhabited their woods. At the same time, the prospect of a permanent, or at least sustained, English presence in the American wilderness raised concerns about what the impact of such a residency would have on the people involved. If, as the English believed, part of their identity as a superior people was due to their ideal environment in England, would colonists take on the characteristics of the land where they lived? In order to avoid physical and cultural degeneration, the English would either have to occupy lands already very similar to their own, or they would have to transform the environment into an English space.[11]

Even before they set foot in America, the English believed that they had a good idea about what kind of environment they would find. Advocates of colonization turned to contemporary scientific beliefs that indicated that America would be an idyllic land for English habitation. Thomas Morton painted an almost mythical picture of the lands awaiting colonists, contending that "the wise Creator of the universal globe, hath placed a golden meane betwixt two extremes . . . betwixt the hot and cold: and every creature . . . within the compass of that golden meane, is made most apt and fit, for man to use." The "golden meane" was a geographic region in which the hot and cold extremes of all other regions were mixed together to create a moderate climate ideal for the English constitution. By simply tracing the lines of latitude across the ocean, they believed, explorers could predict with reasonable accuracy the climate and products of a given land. Based on this reasoning, the English predicted that America would have a temperate climate, just hot enough to grow exotic goods, but not so hot as to be dangerous.[12]

The first reconnaissance reports from Virginia and New England seemed to confirm the latitudinal theory of a "golden meane." Ralph Lane praised the environment of Roanoke effusively, noting that the climate "is so wholesome, yet somewhat tending to heat, as we have not had one sick since we entered into the country; but sundry that came sick, are recovered of long diseases." Lane's assessment of the healthy air and water was an important indication that colonization was possible, but it was his statement about the commodities in the environment that made it desirable. God must have intended the English to settle the land, Lane suggested. How else could it be explained that valuable products were discovered "with very small search, and which do present themselves upon the upper face of the earth?" Unlike the Spanish digging in the silver mines in Mexico and Peru—or, more accurately, the Indians who actually performed the dangerous work—the English would not have to exert much labor in extracting wealth from their colonial possessions. Early documents almost always included a section on "merchantable commodities," where the vast majority of the environmental descriptions can be found. Few Europeans still held out hope that there were large reserves of gold to be found in America, but other products might prove to be equally lucrative, if not as glamorous. English accounts of the reconnaissance voyages read like shopping lists, carefully itemizing everything according to its use and potential value.[13]

Any missing items, such as olives, oranges, and lemons, could easily be added without much trouble. Arthur Barlowe claimed that the land "bringeth forth all things in abundance, as in the first creation, without toil or labor." Other reports claimed that the land had "never been labored with man's hand" and was a "virgin soil." The emphasis on the "virginity" of the soil was not accidental. By naming the land "Virginia" after the allegedly virgin Queen Elizabeth, the English symbolically invested the environment with the idealized qualities of innocence and unfulfilled promise that they intended to nurture. As unused land, the soil would yield incredible bounty since it had not even begun to exhaust the nutrients that crops required. Surely, experienced English farmers could transform the earth through their sophisticated agricultural techniques and would develop such a huge surplus that the rest of the colonists could devote their attention to more profitable pursuits.[14]

Of course, the whole notion of a virgin soil and a wilderness untouched by human hands was demonstrably false, and the same accounts that made these claims also noted that Native Americans had already established extensive agricultural fields there. Thomas Harriot reported that the Algonquian Indians in Virginia relied primarily on "Pagatowr, a kind of grain so called by the inhabitants: the same in the West Indies is called maize." American corn was so easily grown, Harriot claimed, that "one man may prepare and husband so much ground . . . with less than four and twenty hours labor, as shall yield him victual in a large proportion for a twelvemonth." John Brereton claimed that future colonists would be able to eat more familiar food than the native maize, after he and his fellow explorers of New England planted "Wheat Barley Oats, and Pease, which in fourteen days were sprung up nine inches and more." The English significantly underestimated the amount of labor that went into growing American crops and the ease of planting European grains, a miscalculation that would have significant consequences for colonization efforts.[15]

The rhetoric of a virgin land and an untouched wilderness served a clear purpose in these early promotional accounts. By continually referring to the American environment as a wild and untouched land, European accounts figuratively emptied the lands of its native inhabitants. But as the first reconnaissance voyages traveled up and down the eastern coast, searching for a promising spot to establish colonies, the visitors unwittingly initiated a devastating epidemic that would literally empty vast stretches of the land. Thomas Harriot noticed that shortly after the English arrived in Roanoke, the Indians began to die in significant numbers. Harriot claimed that the sickness was not random but was targeted at specific native groups, for "there was no town where we had any subtle device practiced against us, we leaving it unpunished or not revenged . . . but that within a few days after our departure from every such town, the people began to die very fast." A similar process was taking place in New England, where in 1616 European sailors triggered a massive epidemic that, in some Indian groups, killed 80–95 percent of the population. This "marvelous accident," as Harriot called it, was considered to be further evidence that God intended the English to settle the land. After all, God was apparently clearing away the biggest obstacle to English residency. Ironically, the Indian depopulation, which the English initially interpreted as a benefit to their efforts, resulted in an increasing encroachment of the wilderness upon English settlements. Without Indian maintenance of the forests and fields, weeds and wild animals reclaimed areas near European settlements.[16]

European Colonies in the Virginia Wilderness

The English effort to establish a permanent settlement on Roanoke Island failed with the disappearance of the famous Lost Colony of 1587, but advocates of colonization were quick to explain that the failure was not due to any defect of the land. Indeed, the English had identified a more suitable location closer to the Chesapeake Bay. When the Virginia Company decided to renew the English efforts at colonizing the New World, their employees chose Jamestown Island as an ideal spot. John Smith proclaimed, "[H]eaven and earth never agreed better to frame a place for men's habitation being of our constitutions, were it fully manured and inhabited by industrious people." Smith's statement implied that the Native Americans who lived nearby were not applying the same work ethic that English farmers would bring to America, offering a justification for appropriating Indian lands. Robert Johnson's propaganda pamphlet, *Nova Britannia,* argued that Virginia was "inhabited with wild and savage people, that live and lie up and down in troops like herds of deer in a forest: they have no law but nature." Johnson's literary reduction of the Indians to animals wandering in the woods was based on nothing more than fantasy, but it is only an extreme example of a more common English attitude. Unlike the allegedly wild Indians content to live in a wild environment, the English would bring it to its full potential, which meant they would find a way to make a profit.[17]

Imposing order on the wilderness involved two steps: first, the English would have to study the environment of Virginia in order to understand how it differed from England; second, they would use this knowledge to transform the land into a more "English" space. John Smith sought to move away from the exaggerated accounts of the American environment typical of earlier reports. As a commoner surrounded by gentlemen, Smith was particularly keen on establishing his intellectual credentials, and he considered himself a scholar of the American wilderness. Heavily influenced by Francis Bacon's notions of natural history, Smith believed that what mattered most about nature was its utility for humans. His accounts of Virginia (and later New England) dispensed with the notion of a false paradise and argued that it would take considerable labor before the English would make a profit in America. The natural resources were bountiful, Smith claimed, but the real wealth was in fish, timber, and furs, each of which required work to extract.[18]

The English desire to control the environment was heightened by their belief in the correlation between wilderness and identity. One way to determine whether living in Virginia would cause the English to become wild was to scrutinize Indian bodies. After all, they had lived in that environment for years without benefit of the comforts of English housing, clothing, and food. Any future degradation of English bodies should be inscribed on the bodies of the Algonquians. English descriptions of Indian skin color noted that it was somewhat darker than their own, but the difference was slight. The "tawny" or "swarthy" tone, English observers were quick to add, was due to Indian cultural practices of dyeing the skin rather than an environmental influence. English assessments of Indian bodies also noted their remarkably healthy and strong physique; a group of Susquehannock Indians thoroughly impressed the newcomers, since "such great and well proportioned men, are seldom seen, for they seemed like giants to the English." While the prospect of a group of giant men roaming the woods must have given the English some pause, it was a comfort to know that the environment could support, and perhaps even improve, English bodies. But the colonists would have to come to terms with these native inhabitants.[19]

At the same time that the English were busy calculating the material worth of the American environment, the Powhatan Indians were in the process of constructing an empire of their own, based on conquering neighboring Indian groups and exacting tribute. Although there was no sense of private land ownership as Europeans understood it, the Powhatans sought to establish dominion over others in their lands, including the English settlers. The Powhatans had worked for years to create an ideal environment for themselves in Virginia; they knew the best hunting and fishing grounds, they practiced controlled burns to keep down the underbrush, and they had prepared agricultural fields to supply a significant source of food. In short, although the English did not recognize it, the Powhatans had already domesticated much of what the Europeans described as wilderness. After the English had been in Jamestown for a year, it was clear to the Powhatans that their new neighbors intended to become a permanent presence. Tensions rose and a series of minor skirmishes threatened to escalate into a larger war. John Smith met with Chief Powhatan to discuss the matter. "Think you I am so simple not to know," Powhatan reportedly asked, "[that] it is better to eat good meat, lie well, and sleep quietly with my women and children, laugh and be merry with you, have copper, hatchets, or what I want, being your friend; than be forced to fly from all, to lie cold in the woods, feed upon acorns, roots, and such trash and be so hunted by you that I can neither rest, eat, nor sleep?" In his rhetorical question, Powhatan explicitly contrasted symbols of wilderness (cold woods, acorns, and roots) with symbols of civilization (tools, trade goods, and domesticity); clearly Powhatan did not live, nor desire to live, in the wilderness. Although their precise definitions of wilderness may have differed considerably, both the Indians and the English wanted to live in a controlled environment.[20]

The first few years of English occupation were a serious test for even the most optimistic proponents of colonization. Only one month after they built their fort at Jamestown, the colonists started dying of "cruel diseases [such] as swellings, fluxes, [and] burning fevers." John Smith attempted to convince the English to do what he had observed the Indians doing—move seasonally as the local environment changed—but the leaders of the Virginia Company refused to allow the English to mimic what they viewed as savage behavior. Instead, the colonists huddled inside the fort, suffering from typhoid, dysentery, and salt poisoning even though fresh drinking water was only a short distance away. During the infamous "starving time" of 1609, the desperate colonists resorted to cannibalism to survive. Ironically, the English insistence on sedentary living as a sign of civilization caused them to become, at least temporarily, the wild men they feared most.[21]

Indeed, a stunning reversal seemed to have taken place: the English integration plan designed to draw the Indians to "civility" backfired, and the colonists could no longer be certain that they were a distinct and superior people. After learning of an Indian attack on the English in 1622, Smith wrote that "it hath oft amazed me to understand how strangely the Savages hath been taught the use of our arms, and employed in hunting and fowling with our fowling pieces, and our men rooting in the ground about tobacco like swine." In Smith's estimation, the English had degenerated not just to a state of barbarism, but into animalistic behavior. How such a transformation had taken place in only fifteen years of colonization was a matter of great concern. If the causes were external, the settlers could overcome the degeneration by recommitting to maintaining an English space in the New World. If, however, the cultural decay was an indication of an inherent weakness in English identity, degeneracy would remain a constant threat in all colonial encounters.

The Virginia Company decided that the solution was to thoroughly alter the wilderness. All signs of Indian culture were to be removed from the region surrounding the English settlements, transforming them into a more clearly defined English space. Samuel Purchas argued that the dead bodies of the English established a special claim to possession of Virginia: "the dispersed bones of their and their countrymen's since murdered carcasses, have taken a mortal immortal possession, and being dead, speak, proclaim and cry, this our earth is truly English, and therefore this land is justly yours, O English." The living were left to claim the fields that had once been Powhatan hunting grounds and cornfields and that were already bearing the marks of English tobacco plantations. Despite rhetoric advocating genocide, though, elimination of the nearby Indians was out of the question not only because it would remove a necessary source of food, but it might also have unintended consequences for English identity. John Martin reasoned that the surrounding Indians "have ever kept down the wood and slain the wolves, bears, and other beasts (which are in great number)" so that by pursuing a policy of genocide "we shall be more oppressed in short time by their absence, than in their living by us both for our own security as also for our cattle."

The destruction of the Indians would convert English space into Indian space by restoring the wilderness that threatened to envelop the young colony.[22]

European Colonies in the Northern Wilderness

While the English settlers of Jamestown struggled to transform the wilderness into an English space, other colonial advocates argued that they would have better success in the cooler climate of New England. The craggy coastline was not encouraging at first glance. "It is a country rather to affright, then [*sic*] delight one," John Smith reported after his exploratory voyage in 1614, "and how to describe a more plain spectacle of desolation or more barren I know not." An attempt to establish a permanent colony in Sagadahoc, Maine, in 1607 had collapsed when the colonists could not handle the extreme cold. The failure of that group to return with any sort of commodities for the investors was discouraging, but not totally unexpected. After all, one of the chief advocates of colonization argued, "it be not to be looked for, that from a savage wilderness, any great matters of moment can presently be gotten, for it is art, and industry that produceth such things." The first settlers to establish a permanent English presence in the North, however, were not much interested in art, industry, or making a profit.[23]

The thirty-five religious dissenters crammed aboard the overcrowded *Mayflower* in 1620 were part of a failed experiment to establish a religious utopia. These Pilgrims, as they later became known, had grown alarmed at what had happened to their children after only a few years in the Netherlands, where a harsh physical environment and a permissive social one made it likely that "their posterity would be in danger to degenerate and be corrupted." With the parents working night and day merely to survive, there was little time left to devote to spiritual instruction and reflection, the very reason they had left England in the first place. Advocates for establishing a settlement in America pointed out that it was "fruitful and fit for habitation, being devoid of all civil inhabitants, where there are only savage and brutish men which range up and down, little otherwise than wild beasts." Others, perhaps having read about the struggles in Virginia, worried that "the change of air, diet and drinking of water would infect their bodies with sore sicknesses and grievous diseases." Despite the potential drawbacks, the Pilgrims, along with sixty-seven "strangers," landed in Massachusetts in December.[24]

An exploratory party immediately began to search for a likely spot to establish a settlement. Everywhere they went, they passed signs that what appeared from the coast to be a vast unbroken forest was actually a land already transformed by its inhabitants. Woods and meadows showed the telltale evidence of having recently been burned, corn stubble still stood where the ears had been harvested months earlier, and William Bradford found himself dangling upside down, caught in an Indian animal trap. These signs were little comfort to the exhausted settlers who, having selected their site, paused to look around at their new surroundings. The land that the Indians would have considered to be domesticated space, not wilderness, took on an entirely different image in English eyes. Writing many years later, Bradford recalled that "the whole country, full of woods and thickets, represented a wild and savage hue." It was a frightening prospect to establish an idealized society in these environs, but the devout among them reminded themselves that the Bible was full of stories of people lost in the wilderness, only to be redeemed through their faith in God.[25]

The Pilgrims and their fellow colonists set about building their colony, establishing a lifestyle that was not much different from the Native Americans who lived nearby. Much to the dismay of investors back in England, the Plymouth colony never made a profit, and its settlers seemed content to live in poverty. With only a single plow and a few cattle, which did not even arrive until 1624, the Pilgrims did not have the means to even attempt to remake the landscape into an English one. Efforts to grow English barley and peas were largely failures, so the colonists instead planted corn with the help of the Indians. The harvest festival of 1621, which later became mythologized as the first Thanksgiving dinner, was a three-day celebration of American foods; duck, geese, venison, and eel, not turkey, were the likely delicacies on the menu. The major transformation of the wilderness, the efforts that could justly claim to be establishing a "new England" in the wilderness, arrived with the Puritans in 1630.

The Puritans who established the Massachusetts Bay colony were particularly concerned about the possibility of becoming wild in the American wilderness. Only a thin veneer separated humans from beasts, they believed, and one must remain ever vigilant to avoid letting the inner beast take over. In America, the Puritans hoped to establish a "city upon a hill," an example of pure living for the world to emulate. Adam and Eve had been expelled from the Garden of Eden for their disobedience, the ministers preached, and it was the Puritans' intent to get back inside. The church itself was compared to an oasis, in direct contrast to the wilderness of the outside world, where wild animals and wild people roamed. In expanding the church and converting the heathen Indians to Christianity, the Puritans would cultivate new values of human ascendancy over plants and animals as they went. Indeed, John Winthrop argued, "The whole earth is the Lord's garden and he hath given it to the Sons of men with a general commission [to] . . . increase and multiply, and replenish the earth and subdue it." The

Puritans would construct a garden in the American wilderness, but first they had to subdue the outer wilderness and its native inhabitants.[26]

Simultaneously, another group of devoutly religious men was competing for Indian souls in the colder environs of modern-day Canada. The French Jesuit missionaries who arrived in the region in the early seventeenth century described a "desert and barren region, despoiled and desolate of everything." Father Pierre Biard argued that western America was the land of Satan, and "if you consider Satan opposite and coming up from the West to smite us; *A Garden of delight lies before him, behind him a solitary wilderness.*" The French efforts to convert the heathen Indians were made particularly difficult by the untamed environment, Biard claimed, but he went further to explain, "I do not believe that the land, which produces trees as tall and beautiful as ours, will not produce as fine harvests, if it be cultivated. Whence, then, comes such great diversity? Whence such an unequal division of happiness and of misfortune? of garden and of wilderness? of Heaven and of Hell?" The answer, of course, is that to Biard and his fellow Jesuits, as well as to the Protestant reformers in New England, the American wilderness was a battleground of larger spiritual forces pitting the benevolence of God (the garden) against the malevolence of Satan (the wild). While the Jesuits were much more successful than the Puritans in winning Indian converts, their efforts to alter the physical wilderness were much more limited. France did not commit large numbers of colonists to the Canadian forests, content instead to develop a fur trade that spanned the entire continent by the eighteenth century. The depletion of fur-bearing animals had significant social and environmental consequences, but it was left to the New Englanders to create an actual garden.[27]

During the Great Migration from 1630 to 1640, about 14,000 immigrants arrived in New England. The Puritans of Massachusetts Bay immediately set about establishing an English presence in the American wilderness, where they could prosper spiritually and economically. Profit and religion went hand in hand for many New Englanders; the Puritans were careful to avoid becoming worldly, but were not averse to celebrating the fruits of hard labor. Seaports bustled with colonists selling furs, timber, and other goods for distant markets. Churches, towns, and schools were certain signs that the English were winning battles over the Devil and the wilderness, but there were a few suggestions that the war had not yet been won. Wolves seemed to be increasing in number on the outskirts of Puritan settlements, and an alleged outbreak of bestiality cases in 1640–1642 inspired much anxiety about a blurring of lines between humans and animals. In a bizarre scene, Puritan officials brought the suspect animals and the accused men before the courts and summarily executed both after they were found guilty. But just as the Puritans believed that God would ultimately defeat the Devil, they were confident that wilderness would inevitably succumb to domesticity.[28]

First Contact Ends and a Domesticated Wilderness Begins

The introduction of domestic animals throughout America brought a symbolic end to the wilderness during the contact period, causing a significant change in the mental and physical worlds of native peoples and European settlers. In the Spanish-controlled regions, domesticated animals replaced irrigated crops as the primary source of wealth, causing massive environmental destruction. For Indian men, hunting, a significant source of prestige and power, was eroded, as there was little honor in killing a slow-moving cow. Where streams and rocks had vaguely marked hunting territories, fences now crossed the land to protect crops from wayward animals and to demonstrate ownership. Not surprisingly, cattle and fences became prime targets when the Indians of New England launched a last-ditch effort to rid themselves of the English in what became known as King Philip's War. Daniel Gookin tried to prevent the violence, appealing to the colonists that "fighting with Indians about horses and hogs" were "matters too low to shed blood." But horses and hogs stood for all of the changes that had taken place in the American wilderness during the seventeenth century. The Wampanoag Indians and the English settlers both knew that the shape of the land and their very identity rested on the presence or absence of domestic animals and all they represented. Blood *was* shed in New England, Virginia, and Mexico, where the Aztec vision of a gaping wound in the sky proved to be a distressingly accurate premonition of the impending changes in the American wilderness.[29]

In just over a century of sporadic contact between Americans and Europeans, a significant shift took place in both the mental concept of wilderness and the physical landscape itself. In the earliest stage, Spanish explorers struggled to explain the unfamiliar sights and sounds of the circum-Caribbean even as their eyes widened at the sight of towns apparently laden with precious metals. Wilderness was primarily an obstacle to be overcome on the way to securing gold. News of the potential for wealth led to increased competition between European powers during the second stage of contact, when explorers sailed up and down the eastern coast of America and assessed the wilderness as a commodity, moving beyond the early Spanish obsession with gold to identify other material resources. The establishment of permanent colonies marked the final stage of contact and heralded the beginning of a new ideal in America, where wilderness carried negative connotations and beauty rested in fences and farms.

Notes

1. Miguel Leon-Portilla, ed., *The Broken Spears: The Aztec Account of the Conquest of Mexico* (Boston: Beacon, 1962), 4–12.

2. Bernard Sheehan, "Paradise and the Noble Savage in Jeffersonian Thought," *William and Mary Quarterly* 26:3 (July 1969): 328.

3. Keith Thomas, *Religion and the Decline of Magic* (New York: Scribner, 1971), 223, 226; Joyce Chaplin, *Subject Matter* (Cambridge, Mass.: Harvard University Press, 2001), 13; Keith Thomas, *Man and the Natural World* (Oxford: Oxford University Press, 1983), 79; Carolyn Merchant, *Ecological Revolutions: Nature, Gender, and Science in New England* (Chapel Hill: University of North Carolina Press, 1989), 116–17; Gary Nash, "The Image of the Indian in the Southern Colonial Mind," in *The Wild Man Within*, ed. Edward Dudley and Maximillian Novak (Pittsburgh, Pa.: University of Pittsburgh Press, 1972), 56.

4. Bernal Diaz del Castillo, *The Discovery and Conquest of Mexico, 1517–1521*, trans. and ed. A. P. Maudslay (New York: Grove, 1958), 190–91, 216.

5. Miguel Leon-Portilla, *The Aztec Image of Self and Society*, ed. J. Jorge Klor de Alva (Salt Lake City: University of Utah Press, 1992), 154; Thomas Whitmore and B. L. Turner, *Cultivated Landscapes of Middle America on the Eve of Conquest* (Oxford: Oxford University Press, 2001), 220–23; Emily McClung de Tapia, "Prehistoric Agricultural Systems in the Basin of Mexico," in *Imperfect Balance*, ed. David Lentz (New York: Columbia University Press, 2000), 134; William M. Denevan, "The Pristine Myth: The Landscape of the Americas in 1492," *Annals of the Association of American Geographers* 82:3 (1992): 375.

6. Whitmore and Turner, *Cultivated Landscapes of Middle America*, 235.

7. Suzanne K. Fish and Paul Fish, "Prehistoric Environment and Agriculture in the Hohokam of Southern Arizona," in *Native Peoples of the Southwest*, ed. Laurie Weinstein (Westport, Conn.: Bergin and Garvey, 2001), 25–48; Michael Logan, *The Lessening Stream* (Tucson: University of Arizona Press, 2002), 27–30.

8. Stephen J. Pyne, *Fire in America* (Princeton, N.J.: Princeton University Press, 1982), 71–75; Omer Stewart, *Forgotten Fires* (Norman: University of Oklahoma Press, 2002), 3, 25; William Cronon, *Changes in the Land* (New York: Hill and Wang, 1983), 51.

9. Daniel Francis and Toby Morantz, *Partners in Furs* (Montreal: McGill-Queen's University Press, 1989), 6–10; Daniel Richter, *Facing East from Indian Country* (Cambridge, Mass.: Harvard University Press, 2001), 3–5.

10. Margaret T. Hodgen, *Early Anthropology in the Sixteenth and Seventeenth Centuries* (Philadelphia: University of Pennsylvania Press, 1964), 260, 265; Joyce Chaplin, "Natural Philosophy and an Early Racial Idiom in North America: Comparing English and Indian Bodies," *William and Mary Quarterly* 54:1 (1997): 235, 239. See also Peter Mason, *Deconstructing America: Representations of the Other* (New York: Routledge, 1990), 79.

11. William Harrison, *The Description of England*, ed. Georges Edelen (1587; repr. Ithaca: Cornell University Press, 1968), 324.

12. Thomas Morton, *New English Canaan*, in *Tracts and Other Papers*, ed. Peter Force (Gloucester, Mass.: Smith, 1963), 2:11.

13. Ralph Lane, "Letter to Sir Francis Walsingham," in *New American World: A Documentary History of North America to 1612*, ed. David B. Quinn (New York: Arno Press, 1979), 3:189, 290 [henceforth *NAW*]; John Brereton, *A Briefe and True Relation of the Discoverie of the North Part of Virginia, NAW*, 3:350.

14. Arthur Barlowe, *The First Voyage Made to the Coasts of America, NAW*, 3:280; Richard Grenville, "Sir Richard Grenville Reports to Sir Francis Walsingham," *NAW*, 3:293; Kathleen Brown, *Good Wives, Nasty Wenches, and Anxious Patriarchs* (Chapel Hill: University of North Carolina Press, 1996), 15, 22.

15. Thomas Harriot, *A Brief and True Report of... the Countrey of Virginia*, in *The Principal Navigations, Voyages, Traffiques, and Discoveries of the English Nation*, trans. and comp. Richard Hakluyt (Glasgow, Scotland: University of Glasgow Press, 1904), 8:362; Brereton, *A Briefe and True Relation of the Discoverie of the North Part of Virginia*, 349.

16. Harriot, *A Brief and True Report*, 152; Dean Snow and Kim Lanphear, "European Contact and Indian Depopulation in the Northeast: The Timing of the First Epidemics," *Ethnohistory* 35:1 (Winter 1988): 28.

17. John Smith, *A Map of Virginia*, in *The Complete Works of Captain John Smith*, ed. Philip L. Barbour (Chapel Hill: University of North Carolina Press, 1986), 2:111 (hereafter cited as *CWJS*); Robert Johnson, *Nova Britannia, Offering most excellent Fruites by Planting in Virginia,* (London, 1609), C2. In one of his earliest accounts of Virginia, Smith referred to the "vast and wild wilderness" that stood in direct contrast to the Indian towns he also encountered while on an exploration expedition. For Smith, "wilderness" simply meant undomesticated space; the term did not carry with it the same spiritual connotations that Englishmen and women with a more religious world view would have had. See Smith, *A True Relation of... Virginia, CWJS*, 1, 45.

18. James David Taylor, "'Base Commoditie': Natural Resource and Natural History in Smith's *The Generall Historie*," *Environmental History Review* 17 (Winter 1993): 73–89.

19. Smith, *A Map of Virginia*, 149, 162; Gabriel Archer, *The Discription of the now discovered River and Country of Virginia, NAW*, 5:276.

20. John Smith, *Proceedings of the English Colonie in Virginia, CWJS*, 1, 247.

21. George Percy, "A Discourse of the Plantation of... Virginia," *Tyler's Quarterly Historical and Genealogical Magazine* 3 (1922): 272, 273; Carville Earle, "Environment, Disease, and Mortality in Early Virginia," *Journal of Historical Geography* 5 (1979): 391–401.

22. Samuel Purchas, *Hakluytus Posthumus, or Purchas His Pilgrimes* (New York: Macmillan, 1906), 20:228; John Martin, "The Manner Howe to Bringe the Indians into Subjection," in *The Records of the Virginia Company*, ed. Susan Myra Kingsbury (Washington: Government Printing Office, 1903–35), 3:706.

23. Smith, *A Description of New England, CWJS*, 1, 339; Letter, Sir Ferdinando Gorges to Lord Salisbury, 7 February 1608, *NAW*, 3:440.

24. William Bradford, *Of Plymouth Plantation*, ed. Samuel Eliot Morison (New York: Knopf, 1959), 25, 26, 63.

25. Edward Winslow and William Bradford, *Journall of the English Plantation at Plimoth (Mourt's Relation)* (Ann Arbor, Mich.: University Microfilms, 1966), 4–8.

26. John Canup, *Out of the Wilderness* (Middletown, Conn.: Wesleyan University Press, 1990), 46.

27. Letter of Father Pierre Biard to the Very Reverend Father Claude Aquaviva, 26 May 1614, in *The Jesuit Relations and Allied Documents*, ed. Reuben Gold Thwaites (Cleveland, Ohio: Burrows Brothers), 3:5; Pierre Biard, *Relation of New France*, in ibid., 31–33.

28. Cronon, *Changes in the Land*, 75; Canup, *Out of the Wilderness*, 34–46.

29. Elinor Melville, "Environmental and Social Change in the Valle del Mezquital, Mexico," *Comparative Studies in Society and History* 32:1 (January 1990): 28; Daniel Gookin, quoted in Virginia DeJohn Anderson, "King Philip's Herds: Indians, Colonists, and the Problem of Livestock in Early New England," *William and Mary Quarterly* 51:4 (October 1994): 621; Jill Lepore, *The Name of War* (New York: Knopf, 1998), 95.

Three

Religion "Irradiates" the Wilderness

Mark Stoll

I WRITE the WONDERS of the CHRISTIAN RELIGION, flying from the depravations of Europe, to the American Strand; and . . . report the wonderful displays of His infinite Power, Wisdom, Goodness, and Faithfulness, wherewith His Divine Providence hath irradiated an Indian Wilderness.

Cotton Mather, opening words to
Magnalia Christi Americana, 1698

It was no coincidence, the great Puritan divine Cotton Mather believed, that so soon after Christopher Columbus discovered America, Martin Luther set the Protestant Reformation in motion. These two providential events surely demonstrated America's central place in God's plan: wilderness and Protestantism had some common destiny. When the hand of God struck Native Americans down with plague and illness, did he not intend for Protestants to take possession of a wilderness world for his purposes?[1]

As Europeans and colonists like Mather reflected on the opportunities, the meanings, the significance, and the destiny that America represented, they often returned to certain religious themes. Along with the interpretations of wilderness that Melanie Perreault described in the previous chapter, the American wilderness offered a place remote from the "depravations" of Europe, where one could, sinless, begin again and build a perfect society according to God's plan. The English had the greatest success in setting up enduring religious utopias in the wilderness and success as well in implanting in American culture at large ideas about wilderness that would survive the gradual extinction of their original purposes. While such colonizers as the Lords Baltimore in Maryland and

Shaftsbury in Carolina planned nostalgic but chimerical neomedieval social orders, devout and zealous Protestants built durable wilderness utopias inspired by the Bible but also insulated from outsider immigration and isolated from the profitable but communally corrosive temptations of plantation agriculture. Some, like the pietistic German sects, succeeded without leaving a broader legacy, but the New England Puritans left an influential heritage. With entire colonies in their control, they self-consciously developed a theology of nature and wilderness that their descendants promulgated across the United States. The denominations that the Puritan tradition generated—the Congregational (now United Church of Christ), Presbyterian, Unitarian, and American (Northern) Baptist churches, Disciples and Churches of Christ, and others—have fostered a spiritualized wilderness tradition characterized by a uniquely reverent and moralistic love of wild and unpeopled nature. The American sense of "wilderness" as a spiritual and moral resource has flourished due mainly to its deep roots in Reformed Protestant spirituality.

The Wilderness Act of 1964 (discussed in chapter 11) perfectly expressed the American understanding of wilderness as a negative space, without humans or human works. The conception of wilderness upon which this act rested drew from many elements: nostalgia for "wild country to be young in"; the veneration of pioneers, mountain men, and explorers like Lewis and Clark; dismay at the heedless destruction of wildlife and land; antimodernism; and, finally, a view of wilderness originally as the most appropriate place to find and worship God and later as a moral and spiritual resource. This religious aspect of wilderness, even while it evolved under the ever-changing influences of culture, science, and economics, has given it a perpetual appeal in America's famously religious culture and has inspired artists, writers, politicians, and activists to great and important deeds for its defense and preservation.

A religious appreciation for "untrammeled" land that people only visit is unusual in the world and deserves explanation. Morality, ethics, and religious practices almost universally deal with the relationship of humans with each other in society or with the supernatural; they presuppose communities of humans *in* the landscape, not a "community of life" that excludes humans and human works. Pagans protected sacred groves and springs, and medieval Christians filled the landscape with shrines and places of pilgrimage, but they did not protect wilderness from "improvements" like shrines, major religious structures, and enclosures, to say nothing of widespread deforestation and the local extinction of native fauna. Such religions as Christianity, Judaism, and Hinduism have traditions of withdrawal to wild places, usually within walking distance of settlements, but as a rule hermits, mystics, prophets, *sannyasin*, and *sadhus* have left society to avoid worldly distractions and for otherworldly purposes, rather than for the positive spiritual value of this-worldly wilderness itself. American wilderness sees remarkably few contemplative hermits, or *sadhus*, but many hikers with the God-filled prose of John Muir in their backpacks. Muir in fact arose from a particular religious tradition: the Reformed Protestantism of John Calvin and the Puritans. Reformed Protestant preachers so devalued humans and their works, and so praised God and his, that they unintentionally made "untrammeled" earth into sacred ground. Thus, to speak of religion and wilderness (as opposed to religion and nature more generally, or wilderness valued for other reasons) is to speak almost exclusively of the encounter of the British, Dutch, and French Huguenot branches of Reformed Protestantism with the remnants of "Eden" in this world. Indeed, the coincidence of discovery and Reformation that fascinated Mather made this encounter possible—and inevitable.

America, Eden, and the Reformed Churches

Wilderness was much on the minds of the English in the seventeenth century, as their little country founded colony after colony and their compatriots set sail to settle the "Indian wilderness." John Locke famously declared, "[I]n the beginning all the World was *America*." "In the beginning," according to the Bible, humanity lived in Eden. Was God's purpose for America then to send humanity back to the beginning, to Eden—and give Christians a chance to resist the serpent this time? Humankind had a second chance in the American wilderness, and this second Eden might even presage Jesus' second coming and thousand-year reign. The millennium, too, much occupied the minds of the seventeenth-century English. Reformed explorers and colonists tended to see in every green and abundant wilderness a new Eden. As the English explored the American coast and prepared to colonize, they readily envisioned Paradise, particularly in the vicinity of Virginia, where the weather was mild and the land well watered and abundant. *A Briefe and True Report of the New Found Land of Virginia* (1590), which included Theodore de Brys's engravings of John White's paintings of his voyage of 1585, vividly illustrates the Eden in English eyes: its frontispiece depicted not Indians, but Adam and Eve. In a faraway bountiful land, nearly naked men and women lived in apparent ease and with few material goods; surely the English had laid claim to Paradise itself.

More zealous Protestants felt that lying about at ease in Eden only distracted devoted Christians from their responsibility to be the instrument of God's will in the world of sinful humanity. To them, the Bible offered paradigms other than Eden with which to interpret what God intended by placing them in such an environment: wilderness as a place of testing and of providence and tutelage. Exodus and Deuteronomy told how Moses led the Israelites out of bondage in Egypt and into a forty-year sojourn in the wilderness. There, God punished misdeeds and rewarded devotion, fed them miraculously on manna, and brought them the Ten Commandments, before finally allowing them to enter the promised land. Later Hebrew prophets often retired to the wilderness, where, like Elijah, they might hear the "still, small voice" of God (1 Kings 19:12) or through God's providence be fed (1 Kings 17:4–6). John the Baptist, the "one crying in the wilderness" in fulfillment of prophecy, clothed in skins and camel's hair and eating locusts and wild honey (Matthew 3:3–4), emerged from the wilderness to prepare the way for Jesus, who himself "was led by the Spirit into the wilderness" to be "forty days tempted of the devil" (Luke 4:1–13). Wilderness experience, whether of adversity or prosperity, of struggle or ease, had abundant biblical exemplars to give it meaning.

Thus William Bradford, governor of Plymouth Plantation, the first successful explicitly religious English colony, recalled one of the Separatists' purposes in emigrating to America: a "great hope and inward zeal they had of laying some good foundation . . . for the propagating and advancing of the kingdom of Christ in those remote parts of the world." But if the American South was an Eden of mild weather and abundance, the Separatists far to the north in Plymouth in the late autumn of 1620 saw a cold, friendless shore, "a hideous and desolate wilderness, full of wild beasts and wild men," a place of testing where only trust in God could pull them through. Puritans elaborated Bradford's interpretation in much greater detail during their great migration to the Massachusetts Bay Colony a decade later to establish the most theologically minded and well-educated community in the world. They believed themselves to be a literal second Israel whose experience paralleled the biblical Israel and understood that they lived in the last days. Edward Johnson identified New England as the prophesied place "where the Lord will create a new Heaven, and a new Earth in, new Churches, and a new Common-wealth together." Governor John Winthrop wrote a defense of colonization that circulated as Puritans made preparations to embark. First, Puritans must go to America to fulfill their endtime duty "to help on the cominge in of fulnesse of the Gentiles and to rayse a Bulworke against the kingdome of Antichrist." Second, the wilderness was a refuge for the Church against its enemies whence it would return in triumph to Europe, an allusion to the prophecy in Revelations 12 that a woman (the Church) would find refuge in the wilderness from the dragon (the Antichrist). Furthermore, the Puritans argued, "The whole earth is the lords Garden & he hath given it to the sonnes of men, with a generall Condition, Gen:1.28. Increase & multiply, replenish the earth & subdue it." They could obey better by colonizing an empty continent—Indians were apparently of no account—than by allowing it "to lie waste without any improvement" and staying in a crowded country such as England. The Puritans often comforted themselves in the adversity and difficulty of their early years with the assurance that, like the Israelites, they too were exiled in the wilderness for a time. In both cases, God tested his chosen people and, in both cases, sustained them by his providence and chastised them with adversity.[2]

As the years went on, however, Puritans could no longer view their stay in Massachusetts as a temporary sojourn on the way to the promised land or the imminent kingdom of God. They came to realize that perhaps God intended some other purpose, evidently to do his will in the American wilderness, far from the events in Europe. In truth, for most New Englanders, God's will entailed transforming the disordered wilderness into a pleasing pastoral landscape, since after all God had set Adam in Eden "to dress it and keep it," not to lounge idly in the wilderness. Thus, zealous Puritan pioneers begot industrious Yankee farmers (and lumbermen, miners, and builders of towns, railroads, and industry).

Yet Puritans never forgot that they were in the Lord's Garden already. Many biblical passages taught that nature displayed evidence of God's existence, power, goodness, and wisdom. Christians had long taught that God did not reveal his existence and attributes only in a book, the Bible, which only the tiny minority literate in Hebrew and Greek could read; he had also revealed himself in his works, a "book of nature" that even the unlettered could read. Reformed theologians laid great emphasis on these principles. For example, Calvin commented, "I admit, indeed, that the expression, 'Nature is God,' may be piously used, if dictated by a pious mind. . . . In seeking God, the most direct path and the fittest method is . . . to contemplate him in his works, by which he draws near, becomes familiar, and in a manner communicates himself to us."[3] Like Jesus and the prophets, New Englanders often retired to remote or quiet places in the woods or meadows to meditate and come closer to God. Anne Bradstreet's poem "Contemplations" described her solitary meditations on the colorful beauty of a New England autumn sunset. "Rapt were my senses at this delectable view," she wrote. "If so much excellence abide below, / How excellent is he that dwells on high? / Whose power and beauty by his works we know. / Sure he is goodness, wisdom, glory, light, / That hath this under world so richly dight."[4] Puritan minister Jonathan Edwards habitually walked alone along rivers or in woods and fields and would "behold the sweet glory of God in these things"; sometimes he had intense religious experiences there, as when Christ appeared to him alone in the woods in 1737.[5]

Yet the sense that they walked in Eden again also worried Puritans, because Satan would surely tempt the new American Adam to a second Fall, and Paradise would again be lost. Puritans could never forget that human sin had exiled us all from the Garden of Eden. Thus, for example, in Bradstreet's "Contemplations," that the beauty of the landscape reminded her of "Eden fair" led immediately to thoughts of Adam's Fall, and of Eve, with newborn Cain in her lap, who "sighs to think of Paradise / And how she lost her bliss to be more wise."[6] For his part, Edwards's famous "Sinners in the Hands of an Angry God" decried how sinful man misused God's good earth:

> Were it not for the sovereign pleasure of God, the earth would not bear you one moment; for you are a burden to it; the creation groans with you; the creature is made subject to the bondage of your corruption, not willingly, the sun does not willingly shine upon you to give you light to serve sin and Satan; the earth does not willingly yield her increase to satisfy your lusts. . . . God's creatures are good, and were made for men to serve God with, and do not willingly subserve to any other purpose, and groan when they are abused to purposes so directly contrary to their nature and end. And the world would spue you out, were it not for the sovereign hand of him who hath subjected it in hope.[7]

Thus, God charged Adam, Eve, and their descendants to be wise stewards of the earth. Compared to the works of God in nature, the works of man were vain, insignificant, and even dangerous to the soul. Love of self, money, and worldly things led to abuse of the earth—and expulsion from Eden for Adam and Eve and the threat of expulsion or worse for modern mankind. The Puritans had spiritualized wilderness, and there they had seen the will and providence of God, felt his presence, imagined themselves in Eden—and understood well that Satan desired to destroy it the same way he destroyed the first Eden, through the vain, sinful, selfish desires of fallen man.

Wilderness and Other Christian Traditions

Catholic Europeans also imagined an opportunity to reoccupy Paradise, but in contrast to the empty continent of Puritan imagination, their Paradise thronged with Christians newly brought to the faith. Columbus's voyage occurred in an atmosphere of intense crusading zeal in Spain, which in the same year defeated the last Spanish Islamic kingdom and expelled the Jews. During Spain's glorious next century, this heightened religiosity produced the remarkable figures Ignatius Loyola, John of the Cross, and Teresa of Avila. Cardinal Cisneros and other Church leaders imagined the creation of a Christian utopia in

America, far from the corruptions of Europe. Vasco de Quiroga arrived in Mexico hard on the heels of the conquistadors and attempted to establish Christian communities based on Thomas More's recently published *Utopia.*[8] French Catholics also believed that they had a chance to recreate paradise. Affected by the religious revival sweeping France in the 1630s, the directors of the Company of New France hoped to found "a New Jerusalem" in Quebec and attempted to screen the first settlers by their moral qualities. In 1642, a secret radical religious society, the Société de Notre Dame de Montréal, founded the city of Montreal deep in Indian territory as a Catholic utopia for converting Indians to the true faith and ministering to their needs. Unfortunately for their designs, Montreal was also superbly situated for fur trading.[9] In neither the Mexican nor Canadian utopias did Indians play their prescribed roles, and the influx of the Europeans' more worldly and profit-minded compatriots brought these experiments to a quick end.

Catholic spirituality flourished in the un-Edenic wilderness of the present-day American Southwest, where Spaniards found an austere and unforgiving landscape similar to Spain's. Spanish colonization of that remote area attracted zealous, radical Franciscans eager to create a Christian utopia far from European depravations, and, like the English, they were full of millennial expectations. They did not come to imitate St. Francis, the gentle founder of their order, and preach to birds or extol brother sun and sister moon. Instead, Franciscans sought out heathen peoples whom they could convert or at whose hands they might suffer glorious martyrdom, the fate of Fray Juan de Padilla as early as 1542. The dry and rugged Southwest mortified sinful flesh and prepared the soul for heaven, and some sought it out for the adversity that sanctifies. Franciscans drew from the traditional Spanish glorification of patient endurance of pain and suffering for God. Catholic veneration of the saints abounded with imagery of their tortured deaths and faithfulness to the end. Portrayals of the suffering of Jesus and particularly of the sorrows of his mother, Mary, during the events of his crucifixion, confronted worshippers throughout Spanish America, and when Eusebio Kino entered southern Arizona on the first mission to the Indians there, he carried a Mexican painting of *Nuestra Señora de los Dolores* (Our Lady of Sorrows).[10]

Identification with saints and martyrs rather than Adam in Paradise or Moses in the wilderness indicates the vast gap between Catholic and Puritan conceptions of themselves and their relationship to the land and the people in it. In general, Catholics valued wilderness not for its own qualities but as a place without European vices where saintly missionaries could bring the heathen to Christ or die a glorious martyrdom in the attempt. The social genius of Catholic Christianity led Catholics to focus religious attention on people and to reject the individualism inherent in Protestant spirituality, which deemphasized the corporate theology of Catholicism and left the believer alone before God. As the body of believers in which the spirit of God was present, the Church offered saving grace through the seven sacraments. Believers sought salvation and knowledge of God in the Church, not alone in the woods. For a millennium, Catholics had adhered to an Edenic vision expressed in the monastic ideal of a community in the wilderness dressing and keeping the land according to God's command to Adam. Even Eden was cultivated, and so was the land around monasteries and, later, around Franciscan missions from Florida to California. The Catholic ideal landscape has been populated and not wild, the better to benefit the common good, provide for the poor, and promote social justice. Consequently, although Catholic authors, among them E. F. Schumacher, Thomas Berry, Rosemary Radford Ruether, Matthew Fox, and Barry Lopez, have dealt with environmental and nature issues, there have been no Catholic Henry David Thoreaus or John Muirs or David Browers.

Neither did the southern landscape that moved de Brys to include an engraving of Adam and Eve in Paradise inspire a southern wilderness preservation movement or religious tradition. Southerners continued into the eighteenth century to describe their land as an Eden, but metaphorically and not literally as a Paradise to recreate or preserve. Such southern cultural markers as radical individualism, distrust of governmental power, and faith in capitalism impeded the development of wide popular support for the protection of wild landscape. Southern Methodists, like Catholics, developed no wilderness spirituality, while Southern Baptists deeply mistrusted mysticism and preferred God's word in the Bible to his words in the book of nature. Despite the rise of camp meetings after 1800, almost always located in clearings in the woods, few accounts have survived of southerners seeking and experiencing religious experiences alone in the woods and mountains.

Southern blacks similarly thought of "wilderness" and the "promised land" as metaphors full of biblical significance, not as literal places. The Exodus story of Moses leading the Hebrews out of slavery and through the wilderness to the promised land held immense meaning to blacks first under slavery and then under segregation. While wild woods and swamps near southern plantations might mean opportunities to hunt and fish during free time or refuge from abusive masters, in black Protestantism, wilderness was not a destination but a difficult route to a land of milk and honey. Before 1865, blacks yearned for the promised land of northern freedom, after Emancipation for the promised land of cheap homesteads in places like Nicodemus, Kansas, and after 1915 for the promised land of northern cities for their job opportunities and escape from segregation. Recent studies show that inhabited landscapes, with developed parks or open spaces, appeal to African Americans more than do unpeopled wilderness. Of the various religious and secular ways that wilderness has appealed to whites, none has had much resonance for African Americans.

Wilderness and the New England Diaspora

Calvinism faded, but Reformed views of nature and concerns about humans' greedy propensity to ruin the Garden evolved and spread throughout an increasingly secular America. Beginning in the early nineteenth century, a vast migration took New Englanders into unsettled wilderness, across upstate New York, through the Great Lakes region between Ohio and Wisconsin, and on to coastal California and Oregon, where their cultural influence still holds sway. Most moved into the wilderness to set up conventional if pious communities and farmsteads, and they regarded the astonishing abundance of forests and wildlife as God's providence for their use and decidedly not (or not yet) as an Eden to be preserved from sinful abuse.

Mormons, who in the early days almost universally had New England roots, dramatically replayed the Puritan construction of the perfect society in the wilderness. Joseph Smith, the Mormon prophet and a descendant of Puritans, believed that the original Eden had been in Missouri, and he led Mormons there to build God's Kingdom and await the imminent return of Jesus. Chased from Missouri, he and his followers built a utopian community in the wilderness along the banks of the Mississippi, at Nauvoo, Illinois. Rising tensions with non-Mormons led to Smith's murder in 1844 and the Mormons' exodus from Nauvoo in 1846. Mormons revived the Puritan notion of reliving the providential wilderness experience of the ancient Hebrews: the Mormon Moses, Brigham Young, led them across the Mississippi River without getting wet—it froze right before their departure—just as the Israelites had crossed the parted Red Sea. On their journey through the wilderness, providence provided for the Mormons with the miracle of the quail and "manna." They gathered in the promised land, the wilderness of Utah, and began to prepare for the imminent end of times. The redemption of wilderness through labor to make it bloom as it had before Adam's Fall and Noah's Flood would prepare the earth for Christ's expected return. Under church auspices, settlers built irrigation works and loosed their cattle onto the grasslands of the valleys and meadows of the mountains. But overgrazing turned grassy valleys to sage and weeds and eroded mountainsides, from which floods clogged irrigation works and covered fields with sand, silt, and boulders. In the end, Jesus did not come, and Mormons paid an ecological price for irrigation and overgrazing, but for a while they might be excused for thinking they were succeeding in greening a barren land for God. To Mormons, wilderness has been not Eden but a fallen landscape to be redeemed; although they live in the midst of dramatic Western scenery, Mormon advocates for wild nature have been few, notably the Udall family of politicians and author Terry Tempest Williams.

Amid the fervent religious revivalism of the early nineteenth century, the Puritan practice of solitary meditation in woods and fields had fertile results. Like Edwards, Joseph Smith saw visions alone in nature, including one around 1820 in a "silent grove" in which God and Jesus appeared to him;[11] so did numerous others. The New England elite, whose Puritanism had evolved into Unitarianism, preferred now to experience the divine in natural settings. America's leading poet, William Cullen Bryant, declared in "A Forest Hymn" in 1825, "The groves were God's first temples" and were yet the fittest place to worship him, far superior to any work of man. Urban folk without the opportunity to meditate in God's groves purchased landscape paintings, as discussed in chapter 6 by Angela Miller, to bring the moral effects of wilderness into the home. Chief among these religiously motivated landscape painters, almost all of whom had a Reformed or New England connection, Thomas Cole wrote an essay in 1835 that illustrated the painter's implicit Reformed conception of spiritualized wilderness:

> Prophets of old retired into the solitudes of nature to wait the inspiration of heaven. It was on Mount Horeb that Elijah witnessed the mighty wind, the earthquake, and the fire; and heard the "still small voice"—and that voice is yet heard among the mountains! St. John preached in the desert;—the wilderness is yet a fitting place to speak of God. . . . In gazing on the pure creations of the Almighty, he [who looks on nature] feels a calm religious tone steal through his mind, and when he has turned to mingle with his fellow men, the chords which have been struck in that sweet communion cease not to vibrate.[12]

With nature poetry on their shelves and wild landscapes on their walls, tourists who sought out inspiring wilderness sites for communion with the Almighty also brought guidebooks along, such as Boston Unitarian minister Thomas Starr King's popular guide to New England's White Mountains.

This road led to nature mysticism and to wilderness spirituality divorced from theology and in the long term from Christianity. Proliferating, competing denominations caused some to turn away from churches with four walls for God's creedless temples in the woods. As Americans headed to the wilderness to "read" God's book of nature, troubling new doubts emerged about God's other book, the Bible. Geological discoveries challenged Genesis; archaeology and the translation of Egyptian hieroglyphs undermined the trustworthiness of biblical history; and higher criticism from Germany revealed the all-too-human history of the writing of the Bible itself. Unlike the Bible, nature could not be falsified, mistranslated, or misread. In 1836, former Unitarian minister Ralph Waldo Emerson tossed down the theological gauntlet with his essay *Nature*, the manifesto of American transcendentalism. Ritual in human-built churches, he wrote, was merely the dead form of others' religious experiences. To have original experiences of God, he urged the individual into the woods, "these plantations of God," where "the currents of the Universal Being circulate. . . . In the wilderness, I find something more dear and connate than in streets or villages."[13] Emerson's thrilling essays grew in popularity for decades and sent thousands to the wilderness to experience "the currents of the Universal Being."

In the second half of the nineteenth century, the New England diaspora flooded westward to California and carried along the greatest of wilderness pilgrims, John Muir, who is given more attention in chapter 8. Born in Scotland among New Englanders' Reformed cousins, the Presbyterians, raised in sectarian zeal in frontier Wisconsin, then deeply affected by transcendentalism, Muir saw everywhere in the beauty of the Sierra Nevada the creating hand of God, who inscribed his sermons in an ice-carved book of nature, the "glacial manuscripts of God."[14] His writings urged city dwellers into the mountains to "[wash] off sins and cobweb cares of the devil's spinning."[15] His 1890 campaign to protect one of the holiest of God's cathedrals in a wilderness Eden resulted in the creation of Yosemite National Park. Muir's Reformed vocabulary of spiritualized wilderness in large part inspired the national park system and presaged a trend in the twentieth century: Presbyterians, like prophets from the mountains crying repentance, would seize leadership of the spiritual wilderness movement from its sedate, elite New Englander priesthood. With their predilection toward preachiness, censoriousness, and dour moralism, Presbyterians often brought to the wilderness movement their "thus saith the Lord" preaching style and a greater evangelical tenor.

Profit-driven enterprises, which once worried the Puritans, now threatened the last and best of God's first temples and triggered the preacher in Muir. Transcendentalists, painters, photographers, poets, and writers too had mourned how spiritual wilderness daily fell under the juggernaut of selfish greed, which Puritans like Edwards had called sin and transcendentalists like Emerson called the "lowest" use of nature. Cole lamented, "I cannot but express my sorrow that the beauty of the landscapes are quickly passing away—the ravages of the axe are daily increasing. . . . We are still in Eden; the wall that shuts us out of the garden is our own ignorance and folly."[16] By contrast, aghast at proposals to dam one of the natural "cathedrals" in his beloved Yosemite National Park, Muir did not mourn but preached fire and brimstone: "These temple destroyers, devotees of ravaging commercialism, seem to have a perfect contempt for Nature, and, instead of lifting their eyes to the God of the mountains, lift them to the Almighty Dollar. . . . Their arguments are curiously like those of the devil, designed for the destruction of the first garden—so much of the very best Eden fruit going to waste."[17] Unfortunately, his vigorous wilderness evangelism failed, and the "devil" gained the Garden of Hetch Hetchy, but the years-long controversy brought out the Reformed Edenic vocabulary that inspired the idea of *Our National Parks*, as Muir entitled one book.

Wilderness and the Twentieth-Century Reformed Tradition

In the twentieth century, the Reformed tradition continued to serve as the fountainhead of a spiritualized wilderness tradition. Author and activist Sigurd Olson exemplifies the continuity. Olson's father was a minister in the Swedish Baptist church, an offshoot of the American Baptist church, itself a product of eighteenth-century New England. Olson won fame for his fight to keep the Quetico-Superior region an undeveloped wilderness, wrote prolifically, participated in a number of conservation organizations, and led the Wilderness Society in the 1960s. His most popular book, *The Singing Wilderness* (1956), contained the mature development of his spiritual view of nature. He wrote:

> The sun was trembling now on the edge of the ridge. . . . Over all was the silence of the wilderness, that sense of oneness which comes only when there are no distracting sights or sounds. . . . I thought as I sat there of the ancient admonition "Be still and know that I am God," and knew that without stillness there can be no knowing, without divorcement from outside influences man cannot know what spirit means.[18]

The serene Puritan wilderness tradition of transcendentalism, the Hudson River School, and writers like Olson continued in the work of such artists of New England heritage as Marsden Hartley (an Emerson devotee), Arthur Dove, Georgia O'Keefe, and Fairfield Porter, brother of photographer Eliot Porter. Photographer Ansel Adams above all brought forward the ideals of the ethereally beautiful, unpeopled landscape to a new century and a new popular audience and created the iconic images of American wilderness. A descendant of New Englanders, including a devoutly Emersonian father, Adams felt drawn to the Yosemite Valley as "a national shrine." Yet it is a shrine to no particular religion, as Adams rejected religious institutions and creeds for a personal "amorphous sense of deity."[19] In untrammeled wilderness, "the clear realities of Nature seen with the inner eye of the spirit reveal the ultimate echo of God." Compared to the "enormous spiritual and inspirational value" of wilderness areas, "no works of man of any kind [have] consequential value."[20] To Adams, only writers, artists, and photographers had the power to educate the people to protect holy wilderness against the power of profit seeking and material exploitation.[21] Olson's writings and Adams's images embodied the old Reformed values of holy landscape, suspicion of man's works and material motives, spiritual presence, and moralism.

The huge popularity of this genteel, passive New England–style wilderness spirituality has never waned, as witnessed by the huge market for posters, calendars, cards, screensavers, and much else that bear photos by Adams, Porter, and similar photographers. As Muir prefigured, however, the spiritual defense of wilderness increasingly landed in the hands of wilderness prophets whose sense of deity was no less amorphous and unorthodox than Adams's but who learned their spiritual style in the Presbyterian church. The careers of Annie Dillard, Rachel Carson, Robinson Jeffers, Edward Abbey, David Brower, and Dave Foreman (whose contributions to the wilderness movement are chronicled later in this volume) serve to illuminate modern paths from the Presbyterian church into the wilderness. Whether doubts about traditional faith led to faith in Edenic wilderness or the other way around, none of them adhered to orthodox Christianity as adults, but all retained some mixture of appreciation of the spiritual value of wilderness and the evangelical style.

Dillard and Carson continued the Puritan tradition of moral or spiritual meditations on nature. A pious girl who went to Sunday school and church camp and an enthusiastic Emersonian in high school, Dillard wrote her master's thesis on Thoreau. More like Bradstreet than Muir, she won the Pulitzer Prize for *Pilgrim at Tinker Creek* (1974), which described her walks along creeks and through the woods of Virginia's Blue Ridge Mountains and her meditations on God's intent in creating the world. Carson's works illustrate how the Edenic narrative and sermonic style evolved into secular forms that never mentioned God. Granddaughter and niece of Presbyterian ministers and daughter of a stereotypically dour mother, Carson wrote three bestselling books that portrayed the sea as a nearly untouched Eden, too vast for the commercial activities of humans to destroy as they had destroyed so much on land. She secularized the Reformed search for God in nature into the quest for the mystery of existence, as in this passage from *The Edge of the Sea*: "Underlying the beauty of the spectacle [of life] there is meaning and significance. It is the elusiveness of that meaning that haunts us, that sends us again and again into the natural world where the key to the riddle is hidden."[22] The realization that the serpent had again entered sacred precincts sent her into the righteous wrath that informs *Silent Spring.* This powerful book indicted arrogant and venal scientists and chemical companies for the destruction they have unleashed on the natural world, for they left nature's book "unread" and spread "elixirs of death" into the air, land, water, and sea and ultimately into fish, birds, animals, and humans. *Silent Spring* ended like a sermon, offering a choice of two paths: one to salvation and harmony with nature, and one to destruction and ecological desolation.

Jeffers and Abbey took the solitary Reformed love of wilderness and suspicion of man and his works to the edge of misanthropy. Son of a Presbyterian minister, Jeffers wrote brilliant poetry that celebrated nature and a transcendentalistic God and denigrated man and his works as transient and even contemptible. "Hurt Hawks" notoriously stated, "I'd sooner . . . kill a man than a hawk." Brower took a Jeffers line for the name of *Not Man Apart*, the newsletter of Friends of the Earth. Abbey transmuted the Reformed fire-and-brimstone sermon into fiery, cantankerous, if thoroughly secular defenses of wilderness and condemnations of those who would develop it. Although his nature-loving mother was a Presbyterian church music director, Abbey left the church to become, as he once said, "not an atheist, but an earthiest."[23] In books such as *Desert Solitaire*, Abbey celebrated wilderness and attacked "industrial tourism" and the relentless press of commercialization into wild places.

Charismatic environmental leaders Brower and Foreman modernized the Reformed evangelical style of Muir. As a Presbyterian with a devout American Baptist grandmother, doubly baptized by Presbyterian sprinkling and Baptist immersion, Brower read the entire Bible at age eleven. Brower presided over the Sierra Club as it transformed itself into the leading national environmental organization and founded other environmental organizations. His evangelical speaking style reminded author John McPhee of Baptist evangelist Billy Graham. Brower even called his standard speech "the Sermon."[24] If less insistently religious than Muir, Brower periodically referred to the wilderness works of the Creator, as in 1957 when he called for the preservation of "those places where the hand of God has not been obscured by the industry of man."[25] Passionate cofounder in 1980 of the radical environmental group Earth First!, Foreman as a teenager had aspired to be a preacher in Churches of Christ, an offshoot of the Reformed tradition that Presbyterians founded, and for which Muir's father had preached. Foreman, however, evangelized not for Christ but for wilderness. Author Susan Zakin described his evangelical speaking style as "rabble-rousing, foot-stomping, fundamentalist-preacher speechifying" that threw "a monkey-wrench into the works of Demonic Progress."[26] Although Foreman lost his religious beliefs in college, with his moralism and zealous evangelism in defense of wilderness against developers, he as much as Muir and Brower carried the weapons of Reformed Protestantism to battle the serpent of profiteering self-interest in the Garden.

While Muir's explicitly religious style has become unusual among wilderness advocates, young environmental hero Julia Butterfly Hill, daughter of a traveling evangelist, kept the spiritual element of wilderness at the center of her *Legacy of Luna*. The book recounted her protest against the heedless logging of ancient forests by living in the top of a millennium-old redwood tree ("Luna") in northern California for 738 days in 1997–1999. There, she experienced a spiritual awakening: "I suddenly realized that what I was feeling was the love of the Earth, the love of Creation." Hill spoke of the forest in terms that Muir or Bryant or Cole would appreciate:

> Every religion in the world builds shrines, temples, and churches so people can worship and feel connected to creation and the Creator. Yet the ancient forest cathedrals are continuously desecrated by industrial logging practices. Protecting the sacred forest ecosystems is a moral imperative on behalf of all life and compels all spiritual people to unite in this common goal.[27]

If few leading wilderness advocates today write so explicitly of the spiritual or religious significance of wilderness, the popular connection of religion and wilderness continues to thrive in less visible ways, particularly as manifested in another form of the Reformed legacy, youth camps. After the Civil War, a wave of organized camping swept New England and led to the establishment of permanent camps for youth and retreats for adults. By 1924, an estimated 90 percent of the country's 1,248 camps were in New England, and they were generally grounded in some moral or spiritual purpose. The Reverend George W. Hinckley founded the first church camp at Gardners Island, Rhode Island, in 1880. Son of devout Presbyterians, naturalist and writer Ernest Thompson Seton founded Woodcraft Indians in 1902 and the Boy Scouts of America in 1910, both movements centered on camps and camping in the wilderness. The Girl Scouts and other organizations soon followed the men's lead and set up camps for girls. The Young Men's Christian Association (YMCA), whose early participants included Muir and Gifford Pinchot, took an early lead in establishing youth camps. In 1910, the YMCA built its most spectacular camp, YMCA of the Rockies, as a conference center and family camp adjacent to Rocky Mountain National Park, Colorado. In one example of the camps' influence, Gary Snyder, the Pulitzer Prize–winning poet whose work swings between meditations on wild nature and disparagements of materialism and greed, was raised without religion but as a youth spent summers at YMCA Camp Loowit, Oregon. A friend recalled:

> On one of our many three- or four-day hikes, I remember that our trip leader—the camp's inspiring founder, white-haired J. C. Mechan himself—while taking a break on the summit of Mount Margaret, looked out over the panorama of the Cascade Range and asked, "What do you see, boys? What do you see?" Slowly, there were scattered replies. . . . Finally, after

some moments of silence, I heard Gary say quietly, "It's God, it's all God."[28]

After World War II, churches of many denominations founded so many camps that yearly stays at summer camps became a common experience for baby boomers, giving them a religious wilderness experience that may well have increased their receptivity to wilderness preservation and recreation and may possibly continue to pass on these values in some form to new generations. Religious camps and retreats now dot the wild landscape. One of the best known is Ghost Ranch, near Abiquiu, New Mexico, which was given by a local rancher to the Presbyterian church in 1955 and which surrounded Georgia O'Keefe's house, and where many still sojourn to experience God in the wilderness.

The American spiritualized wilderness tradition has an unclear future. Always a minority religious tradition in America, Reformed Protestantism has lost "market share" for two centuries and membership for decades. Wilderness figures from Emerson to Muir to Foreman and a disproportionate number of environmental leaders grew up in these churches. Most left them. Some, like Gary Snyder, sought in Asian or Native American religions equivalent connections between spirituality and nature, in essence replacing the waning spiritual legacy of unlovely Calvinism with the putative authenticity of more ancient, gentler traditions. It remains a significant question whether society can transmit spiritual wilderness values to future generations in syncretic or cultural contexts, without denominational institutional structure and moral mandate. Indeed, the number of young people seeking out wilderness is declining: between 1965 and 1995 the average age of a backcountry backpacker increased from mid-twenties to late thirties.[29] After a hiatus of nearly two decades, I recently backpacked the Yosemite backcountry with my son, and the large number of fellow graying baby boomers on the trails compared to the number of young people astonished me, quite the opposite of the scene twenty-five years ago. Many factors have contributed to that development, but the parallel waning of the Reformed tradition may be no coincidence. The growing Protestant denominations, including Pentecostal, Southern Baptist, and Mormon, propagate few doctrines productive of wilderness spirituality. Has the nation seen the last Dillard, the last Brower, the last Foreman?

Cotton Mather was right. When Europeans settled in America, their God irradiated the Indian wilderness. Particularly for children of the Reformed tradition, wilderness has been irradiated with spiritual meaning ever since. While they may love and defend wilderness for other reasons, Catholics, Methodists, Southern Baptists, Pentecostals, Jews, and members of other non-Puritan denominations historically have rarely fostered a similar investment of uninhabited wilderness with spiritual and moral meaning. To Puritans, however, the ideal of doing God's will surrounded by his Creation lay near the center of their purpose and self-conception and left a lasting imprint on American culture, from summer camps to the Ansel Adams calendars on our walls. Particularly among the formerly Puritan Congregational and Presbyterian churches and their denominational offspring, the Unitarian, American Baptist, and "Christian" churches, the old search for God in nature persists. From this Reformed tradition, in both religious and secular contexts, continues to spring a love of wild nature as a source for values or religion, opposition to the greedy commercialization of wilderness and resources, and self-righteous anger and fiery denunciation of the "temple destroyers." Imbued with a sense of the sacredness of wilderness, the sons and daughters of the Reformed tradition have stood ready at the gate of every remaining wilderness Eden, vigilant lest the serpent once again gain entrance. But the devil never rests, and the watch grows weary.

Notes

1. Cotton Mather, *Magnalia Christi Americana*, 2d ed. (Hartford, Conn.: Silas Andrus, 1820), 1:44, 72, 78.

2. William Bradford, *Of Plimouth Plantation*, ed. Samuel Eliot Morison (New York: Modern Library-Random, 1952), 25, 62; Edward Johnson, "Wonder-Working Providence of Sions Saviour in New-England," in *Wonder-Working Providence of Sions Saviour in New-England (1654) and Good News from New England (1648)* (Delmar, N.Y.: Scholars' Facsimiles & Reprints, 1974), 3; John Winthrop, *Winthrop Papers*, ed. Stewart Mitchell (Boston: Massachusetts Historical Society, 1931), 2:115.

3. John Calvin, *Institutes of the Christian Religion*, trans. Henry Beveridge (Grand Rapids, Mich.: Eerdmans, 1964), 1:v5, v9.

4. Anne Bradstreet, "Contemplations," in *The Complete Works of Anne Bradstreet*, ed. Joseph R. McElrath, Jr., and Allan P. Robb (Boston: Twayne, 1981), stanzas 1 and 2.

5. Jonathan Edwards, *The Works of President Edwards*, 8th ed. (New York: Leavitt & Allen, 1856), 16, 21–22.

6. Bradstreet, "Contemplations," stanzas 11 and 12.

7. Jonathan Edwards, "Sinners in the Hands of an Angry God," in *Jonathan Edwards: Representative Selections with Introduction, Bibliography, and Notes*, ed. Clarence H. Faust and Thomas H. Johnson (New York: American Book, 1935), 162–63. Edwards alludes to Romans 8:19–22 and Leviticus 18:28 and 20:22.

8. Charles Gibson, *Spain in America* (New York: Harper & Row, 1966), 69–75.

9. W. J. Eccles, *France in America* (New York: Harper & Row, 1972), 39–40, 47–49; and Kathleen Jenkins, *Montreal: Island City of the St. Lawrence* (Garden City, N.Y.: Doubleday, 1966).

10. See especially Belden Lane, *Landscapes of the Sacred*, 2d ed. (Baltimore, Md.: Johns Hopkins University Press, 2001), 100–113, on whose insights this paragraph is primarily based; also Ramón A. Gutiérrez, *When Jesus Came, the Corn Mothers Went Away: Marriage, Sexuality, and Power in New Mexico, 1500–1846* (Stanford, Calif.: Stanford University Press, 1991), ch. 2.

11. Dean C. Jessee, *Personal Writings of Joseph Smith* (Salt Lake City, Utah: Deseret, 1984), 75.

12. Thomas Cole, "Essay on American Scenery," in *American Art, 1700–1960: Sources and Documents*, ed. John W. McCoubrey (Englewood Cliffs, N.J.: Prentice-Hall, 1965), 99–100, 102.

13. Ralph Waldo Emerson, *Nature*, in *The Prose Works of Ralph Waldo Emerson*, rev. ed. (Boston: Fields, Osgood, 1870), 1:5–9.

14. William Frederic Badè, *Life and Letters of John Muir* (Boston: Houghton-Mifflin, 1924), 1:358.

15. John Muir, *Our National Parks* (Boston: Houghton-Mifflin, 1901), 2.

16. Cole, "Essay on American Scenery," 102, 109–10.

17. Muir, *The Yosemite* (1914; rpt., San Francisco, Calif.: Sierra Club Books, 1988), 196–97, 195.

18. Sigurd Olson, *The Singing Wilderness* (New York: Knopf, 1956), 130–31.

19. Ansel Adams and Mary Street Alinder, *Ansel Adams: An Autobiography* (Boston: Little, Brown, 1985), 59, 18.

20. Adams, Letter to Harold Bradley, Richard Leonard, and David Brower, July 27, 1957, in *Ansel Adams: Letters and Images, 1916–1984*, ed. Mary Street Alinder and Andrea Gray Stillman (New York: Little, Brown, 1988), 248.

21. Ansel Adams, "The Meaning of the National Parks," in *Ansel Adams: Our National Parks*, ed. Andrea G. Stillman and William A. Turnage (Boston: Little, Brown, 1992), 16.

22. Rachel Carson, *The Edge of the Sea*, in Rachel Carson, *The Sea: The Sea around Us; Under the Sea-Wind; The Edge of the Sea* (London: Paladin, 1968), 404.

23. Edward Abbey, *Desert Solitaire: A Season in the Wilderness* (New York: McGraw-Hill, 1968), 163.

24. John McPhee, *Encounters with the Archdruid* (New York: Farrar, Straus, and Giroux, 1971), 79–87.

25. David Brower, *For Earth's Sake: The Life and Times of David Brower* (Salt Lake City, Utah: Peregrine Smith, 1990), 3, 19–20, 131, 134; quotation from "Wildlands in Our Civilization," *Sierra Club Bulletin*, June 1957, quoted on 259. I can find no specific statement of his religious views in his autobiographical writings, other than this: "I am not irreligious, but I do think that many of the religions I know of are artifacts," in Brower, *Work in Progress* (Salt Lake City, Utah: Peregrine Smith, 1991), 272. That statement resembles Emerson's rejection of established religions in nature.

26. Susan Zakin, *Coyotes and Town Dogs: Earth First! and the Environmental Movement* (New York: Viking, 1993), 9.

27. Julia Butterfly Hill, *The Legacy of Luna: The Story of a Tree, a Woman and the Struggle to Save the Redwoods* (San Francisco, Calif.: Harper San Francisco, 2000), 67;

Hill, "Julia Joining the Religious Campaign for Forest Conservation," *Circle of Life* Web site, http://www.circleoflife.org/inspiration/julia/words (accessed August 10, 2004).

28. Jerry Crandall, "Mountaineers Are Always Free," in *Gary Snyder: Dimensions of a Life*, ed. John Halper (San Francisco, Calif.: Sierra Club Books, 1991), 4.

29. David N. Cole, "Wilderness Recreation Use Trends, 1965 through 1994," U.S.D.A. Forest Service Research Paper INT-RP-488, 1996.

Four

Farm against Forest

Steven Stoll

American expansion was agricultural expansion, and it came at the expense of Indians and wilderness. The rise of the United States as a continental power is often understood militarily and ideologically: as the cavalry fighting the Sioux on the Great Plains, as the conflict within Congress over the balance between free and slave states, or as the belief in the nation's "manifest destiny" to span coast to coast. But from the backwoods of Pennsylvania to the Willamette Valley of Oregon, the principal way that the United States took possession of the lands it conquered was through the migration of farmers, who worked forests and prairies into habitable, taxable, defensible space. The politics of land has a material foundation in the relationship between farmers and land.

This chapter dwells on the United States in the nineteenth century to consider the tangled meanings that Americans gave to agricultural expansion through such subjects as the interior policies of the antebellum presidents, the land-use practices of farmers and planters, and the first conservationist thinkers. The confrontation between farmland and forest, however, turns out to be much older than the United States, even older than the first English settlements in North America at Plymouth, Massachusetts, and James Town, Virginia. The shifting line between farmland and wilderness after the American Revolution came from deep within the material world of agrarian peoples.

Ten thousand years ago, people all over the world began to select wild grasses for planting in a long transition with hunting that resulted in settled agrarian societies. In a parallel with the way God separated light from darkness and Eden from the rest of the earth, humans began to divide the landscape between the places they cultivated and those they did not. Hints of the partition that resulted in countryside and wilderness show up in the first works of written English. The word *field* appeared in 1000, defined as a place where flowers or weeds grew as a consequence of human disturbance or intention. It is a clearing, an opening in the woods. *Forest* dates from 1300, referring to that which is found "outside"—outside the field, the farm, the country. Long before the first Puritan settlers landed in 1620, English-speaking people possessed a lexicon that helped them to carve out mental as well as physical spaces, consisting of words that set them off from the environments and the people who lived outside the timbered ramparts of the tiny village on the shores of Cape Cod Bay.

According to anthropologist Hugh Brody, it is a mistake to think of hunters and gatherers as traveling great distances for food while farmers stayed put to build permanent homes. The opposite is true. Farmers have served as the shock troops of the earth's environmental transformation. There are two main reasons for this, and together they form the basis of the feverish land hunger of agrarian societies. Farmers thrive by transferring nutrient elements from soils to domesticated plants and animals. Cultivated soils need some form of restoration—whether from lying fallow or intensive fertilizer—or they run a deficit and decline in productivity. In regions where labor is scarce and land is plenty (a description of the backwoods of North America throughout most of the nineteenth century), farmers had little incentive and even less capital to invest in land. The nearly constant scarcity of nutrients like nitrogen forced them to seek out new land. Agrarian societies also generated people—many more than hunting and gathering societies did—so they continually created their own necessity for expansion, as sons and daughters sought to reproduce the material world of their parents. Swards and scythes really served the same purpose—by geographical or ecological conquest, they turned contested territory into land.[1]

For the first English colonists in North America, as for all subsequent settlers of European lineage, creating the kinds of places they considered to be habitable meant replacing forest with farmland. Yet the questions and the causes were always more complicated than this formulation suggests. The Wampanoag Indians also created openings where they planted corn, beans, and squash in gardens almost identical to those of the English. Standing in fields of maize at Plymouth in 1627, the expansionary trajectory of the one people at the expense of the other would not have been obvious. The farmers who settled Concord, Massachusetts, managed a region of river meadows, pastures, and fields without demanding additional land for two hundred years. The first signs of stress came as the town's population increased in the eighteenth century, and though the ecological balance among all of its elements remained in tension, Concord sent its sons and daughters to cut and clear other places, to establish other towns. Townspeople increased the size of their arable land at the expense of both meadow and woodland by the nineteenth century.[2] In essence, even as Concord seemed to represent constancy and socially acknowledged ecological limits, it finally yielded to the pressures endemic within agrarian societies.

Agriculture's tendency to absorb territory cannot be considered in isolation from capitalism, in which land and labor carry the burden of generating not just subsistence but surplus value, or profit, in exchange. We run the risk of thinking that we are looking through the lens of agriculture when what we are really seeing is the feverish accumulation of capitalism. Combined, the two became a fearsome force. And yet, as we will see, movements within capitalist agriculture itself briefly favored a restorative husbandry designed to prevent migration and to maintain balance in the landscape between field and forest. So was agriculture itself the cause of expansion into the American forest, as Brody argues? Or was it European thinking about property and civilization in the seventeenth century, expressed through capitalism? What about population as the driving force behind migration and the intensification of agriculture? They each became factors in different circumstances, but the pattern is abundantly clear, regardless of the exceptions.[3]

These were the circumstances for the explosion of the countryside all over the world during the seventeenth century: from the Guizhou province of China in the seventeenth century, where the state encouraged internal colonization against the indigenous Miao people, to the Russian peasants who brought their grain-and-hay culture to the newly conquered Kazan frontier at about the same time, to the Europeans who soon ventured out from Plymouth and James Town, New Amsterdam and Santa Fe into the interior of North America. In each of these cases, agrarian people sought land to reproduce their social and material lives, to provide meadows and pastures for their animals, and to establish farms large enough to divide among their sons. The politics of land in North America arose atop these deep structures in the ways and means of agrarian people.[4]

A word on usage is necessary before proceeding, especially given the subject of this book. Though settlers entered regions that we would call *wilderness*, they almost never used that term, and their use of it carried none of the associations we give it—of places valued for the relative absence of a human presence, places worthy of legal protection from economic activity. Until the twentieth century, wilderness served as a relative, not an absolute, category. It defined places and times when humans did not yet control their environment or where they had lost that control. When the scrub and brush of southern "oldfield" fallow turned to woods, planters said that the wilderness had returned. More often, farmers referred to forests and woods, which could mean wartorn frontiers, woodlots, or the backwoods—where families created clearings with fire. When it came to forested environments, in other words, agrarian people recognized a continuum of human influences. For the purposes of this chapter, I will refer to the *woods* and the *forest* to mean the environments where agrarian people converted the eastern wilderness into an agricultural landscape.

One event brings this transformation into relief. It is no accident that the person who closed the biggest real estate deal in human history—making possible the transition of vast woods into countryside—was also the most famous agrarian of the century and a slaveholding southern planter.

Thomas Jefferson and the Louisiana Purchase

When Thomas Jefferson sent two army spies on cross-continental reconnaissance, he had in mind the acquisition of most or all of that territory for the United States. The Louisiana Purchase opened a frontier for settlement so immense and came at a time so significant that it destabilized the balance of power in Congress and led to four decades of state formation and political conflict, bringing on the Civil War. It became the center of the United States, the scene of its most violent and protracted Indian wars, and its preeminent settlers' country. It is also a crystalline example of how government conceived of the continent as a space that would be brought into the political realm through the extension of agriculture. Jefferson himself spoke through metaphor, deploying agrarian ideals to bring about political realities. Here is what he said in the few lines from *Notes on the State of Virginia* that earned him the title of agrarian philosopher: "Those who labour in the earth are the chosen people of God, if ever he had a chosen people, whose breasts he has made his peculiar deposit for substantial and genuine virtue. It is the focus in which he keeps alive that sacred fire, which otherwise might escape from the face of the earth."[5]

About whom was Jefferson writing? What kind of agriculture and where? Jefferson believed that open land and the act of working it had a civilizing effect, that it shaped people by providing them with the material means for subsistence, taught them to defend themselves, and thus nurtured them as citizens of a democratic society.[6] Such unmediated contact with land was exactly what southern planters lacked, since they rarely even supervised their plantations. So Jefferson constructed an ideal (or borrowed one): the freeholder, the yeoman. The yeoman was a substantial English farmer—neither lord nor peasant—who cultivated his own land. Paying no regard to the fragile and declining position of the yeoman class within English landed society at the time, Jefferson claimed it as an American identity, consisting of people who owed their living to no one, guarded their independence fiercely, and moved where they pleased. Jefferson made more than a turn of phrase when he defined them by the fire they carried. From his perch at Monticello, he saw the steady train of families and the glow of their encampments at night. The yeomen of Jefferson's imagination carried real torches over the Blue Ridge, and they set the forest ablaze.[7]

Jefferson observed the people of the backwoods, a settlement culture that possessed such a startling capacity to create openings that nothing but the geological contours of the continent stopped them. Their ax/fire/rifle/log-cabin tradition took its rise in the lower Delaware River valley early in the eighteenth century. Its key characteristics included a classless society, familiarity with and even acceptance of Indians, a subsistence that depended as much on hunting as it did on agriculture, scattered homesteads rather than villages, and constant, compulsive movement—with families moving as often as once a generation or more. These are the people whose notch-log structures can be traced through Pennsylvania and Kentucky, who reached Illinois before 1815, Missouri at about the same time, who broke the sod of the Great Plains in advance of the surveys, and who arrived in Oregon by the 1840s. They hunted out predators and built their homes as squatters in American-, Spanish-, or British-controlled territory.[8] In this way, they carried American influence by force of numbers and thus served as the unacknowledged soldiers of the interior policies of Andrew Jackson and James K. Polk, to name just two presidents following Jefferson who depended on them. Put another way, backwoods people cleared more land more quickly than any other people in human history, and no military or political force ever turned them back. Their needs coincided with Jefferson's own for a robust space for American expansion.

Backwoods settlers and Indians used fire to reduce forest undergrowth to ashes, then planted corn and grazed their animals among charred logs and half-burned trees—a method of shifting cultivation called *swidden*. The destructiveness of this practice is an illusion. The forest comes back to be cut and fired again. Swidden agriculture uses the dynamics of the forest itself to cycle nutrients. Swidden societies all over the world, like those of Southeast Asia, have remained in limited territories for centuries, allowing forests to regenerate and mature before repeating the burn.[9] Yet a sustained swidden cycle never became well established in North America once white farmers took over land from Indians. Even if land did lay fallow for a time, it soon came into the hands of farmers who cultivated it permanently, who turned smoky woods into established agricultural landscapes by the middle of the nineteenth century. Backwoods people conducted a form of agriculture that existed within and depended upon forest dynamics, but in so doing they prepared the ground for farmers whose land use did away with the forest altogether. Once the population of rural districts reached a certain density, swidden threatened private property and became impractical if not illegal.

Jefferson's admiration of backwoods farmers had more to do with the economic equality they represented and less with the methods they used to turn land into agricultural space. He understood that American democracy depended on equal access to land and its resources. The great woodland looms up behind the words "all men are created equal," because in order for that phrase to have had meaning, settlers needed to get hold of the conditions of subsistence with little or no capital, with no resistance from government, with no barriers to their movement, and with no one else laying claim to the land. Jefferson not only acquired territory for this purpose, he authored the Ordinance of 1785, one of three legal instruments for disposing of it. The ordinance founded the survey, or the project of tracing lines across the country in units of six square miles, each broken into thirty-six sections of 640 acres.[10] So, at the same time that backwoods settlers burned, planted, and hunted to change forests into farms, the United States extended its legal boundaries over the same land, changing it into real estate.

No other law spoke as loudly in favor of agriculture as the interior policy of the United States. The survey made it possible for someone to visit a land office in Philadelphia or Richmond and buy or sell land anywhere the survey reached without ever seeing it. The survey intensified military conflict against Indians because the tendency of Indians to hunt and plant over extensive areas and their refusal to acknowledge an invisible geometry of square spaces challenged the logic and the legality of expansion. Though portions of the United States would not be developed for a century or more, and though the most violent confrontations between Indians and the United States lay in the future, the assertion of the grid in the 1780s provided a kind of control over the backcountry. Six-mile squares covering the map of the Northwest Territory worked a mental conquest that continually enabled an ecological one. The settler culture moving in and the tools and institutions of government designed to acquire land came to bear on the Louisiana Purchase.

Farmland Triumphant

The survey was a totalizing idea. It proposed to turn the entire continent into real estate, to give everything exchange value. According to one facet of modern thought, that triumph needed to be complete. This view never had a more remarkable advocate than John Adolphus Etzler, a German engineer who immigrated to the United States in 1831 and set out to establish a community of philosophical and material freedom on the prairies of the upper Midwest. Two years later, he published a kind of handbook on the perfection of nature and society called *The Paradise within the Reach of All Men, without Labor, by Powers of Nature and Machinery* (1833):

> Fellow-Men! I promise to show the means for creating a paradise within ten years, where every thing desirable for human life may be had for every man in superabundance, without labor, without pay; where the whole face of nature is changed into the most beautiful form of which it be capable. . . . [Man] may level mountains, sink valleys, create lakes, drain lakes and swamps, intersect every-where the land with beautiful canals . . . he may provide himself with means unheard of.

By harnessing wind and waves, said Etzler, humans could create an infinitely extendable countryside, destroy every square mile of wild nature, and sustain a human population of one trillion in the greatest imaginable comfort. Addressing Americans, he offered a settler's Eden:

> Your hideous wilderness, that is now but the habitation of brutes and venomous or loathsome vermin and a few scattered miserable Indians, will rapidly become the delightful abodes of happy intelligent human beings. By a simple application of the new means, the soil, so prepared, will be covered with luxuriant growth of all desirable vegetables that the climate admits of, the finest gardens extending many miles in every direction.... Snakes, mosketoes, and other troublesome vermin will have disappeared, the causes of their existence being annihilated.[11]

Like Jefferson, Etzler believed that he articulated universal desires for wealth. Like Jefferson, he believed that virtue is secured by one's material condition. And, like Jefferson, he invented machines. Etzler called his masterwork the "Satellite." It was a kind of multipurpose plow designed to perform all of the functions of land management, powered entirely by wind. It would plow, pulverize and sift soil, level, sow grain, pull weeds and dress soil, cultivate between plants, mow, harvest, hammer, saw, cut down trees, pull out stumps, notch rocks, excavate and elevate, dig ditches, form terraces, operate in water or mud, and dig mines. The Satellite would create spaces recognizable as those belonging to an agrarian society, and its infinite source of energy would ensure the limitlessness of its reach.

The three-ton Satellite failed its two trials—one in Pennsylvania and the other in England—between 1843 and 1845. It would be a mistake to dismiss it and its inventor as farfetched. At nearly the same time, inventors no less visionary built machines remarkably similar in function, if not in breadth, to Etzler's.[12] Steam-driven combines—machines that harvested, threshed, and sacked wheat—created some of the largest cultivated spaces in human history, turning tens of thousands of acres in the Red River valley of South Dakota and the Central Valley of California into wheat fields. The survey and the grid it made, the stripped and plowed paradise promised by the Satellite, wheat fields the size of counties made possible by the steam combine, and capitalism itself represented a totalizing vision of agrarian society.

The forest that once covered the eastern United States from Maine to Florida and to the edge of the Great Plains fell under this regime. American settlers cleared 460,000 square kilometers between 1650 and 1850, and then an additional 800,000 from 1850 to 1910. During the 1870s alone, 200,000 square kilometers fell to fire or ax.[13] The countryside has not covered the earth, as Etzler dreamed, but cultivated land now runs up the sides of the Virunga Mountains in Rwanda and fills in the distances between cities and game preserves throughout Africa. It reaches to within miles of the Arctic Circle and nearly to Tierra del Fuego.

Agrarian Conservationists

Not everyone in the nineteenth century cheered the migrants in their furious clearing of the West. Not all farmers held a totalizing vision of farm against forest. Eastern farmers and planters with ties to communities in the old Atlantic states regarded backwoods clearings as signs of moral failure and harbingers of their own political marginality, as upstart territories gained population and siphoned off power in Congress from the eastern establishment.[14] In response, they devised a form of agriculture (or borrowed it from the English landlords and tenants who had perfected it over the previous three centuries) intended to keep households settled in one place for longer than three generations.

Convertible husbandry, or the practice of converting the same land from arable fields to grass in rotation and of collecting the manure of cattle to cycle nutrients through soils, proposed an agricultural solution to a political problem. The hoped-for result would be a republic that did not expand into Louisiana, one that would hold its ground in order to establish communities of permanence. As one Maryland farmer meditated:

> If a wall, like that of China had been built around "the old thirteen" at the time when they resolved to set up for themselves, how different would be their aspect, and how much more highly cultivated, populous, strong and comfortable at this time—But our policy has been, by the prodigal management of our public domain, to set in motion a constant current of emigration . . . which has drained the old states of their most active and vigorous population.[15]

No one who held such a view would have advocated the complete destruction of wilderness land. Improving farmers did not innovate the modern idea of wilderness. They demonstrated little of the romantic sensibility gaining in popularity among intellectuals. However, by favoring a concentrated population and intensive methods, they made room for wilderness.

Some farmers said that if a restorative husbandry quelled migration, then it might create a landscape of balance and diversity, where land would be spared for woods and waters. Jesse Buel, a New York politician and editor who spoke for northern improvers, worried that the forces of expansion and migration would make every place the same as every other:

> In our zeal to *clear up*, we generally carry the matter to an unwarrantable extreme; every thing is cut away—the whole surface is denuded—stripped of its natural growth. We know that old forest-trees will not long bear an open exposure—that the winds will prostrate them when deprived of the protection of surrounding forests. . . . The settler upon new lands may *preserve*, without labor or expense . . . that which imparts to old-settled districts the highest rural charms, and gives to them much of their intrinsic value. To destroy, in this case, is but the labor of a day; to restore, is the work of an age.[16]

Buel described an American *Landschaft*. The word can be translated as province, district, region, or countryside. It posits a continuum, with a core of nestled homes and gardens at the center, surrounded by a border of fields, then pasture, then forested land, with wilderness more distant still and less immediate. The idyll is one of constancy, in which communities grow ancient without deterioration, in which people never demand more from their environment than it can give. But it never fit American conditions because it could not account for the environmental conditions and economic impulses that drove farmers to wear out the land they owned and seek more. As compelling as agricultural improvement seemed to those who espoused it, they adopted it out of interests and motives that most other American farmers did not have.

No one knew the conditions and impulses that drove migration better than the planters of the southern states. Though the plantation did not travel as far or as wide as the backwoods settlement, it functioned just as well as a cyclone of resource extraction and woodland destruction. It landed in the West Indies well practiced, if not perfected, as a frontier institution after centuries in the Mediterranean and the islands off the coast of Africa. Its first and most enduring form featured slave labor and great measures of land in order to produce cane sugar, a crop Europeans carried to the Caribbean in the sixteenth century. By the nineteenth century, the plantation of the American South produced a wider variety of products, including indigo, tobacco, rice, and cotton, but land enough to feed the punishing demands that cotton placed on soils had become scarce in Virginia by the 1790s and in the Carolinas by the 1820s.

The reason had to do with the way planters thought about land and labor. The cost of human chattel relative to land meant that planters kept the productivity of their slaves ever in mind. Labor—even more than land—needed to be fully employed all the time. Fertile soil made slave labor more productive. Labor expended on newly cleared ground yielded more cotton and corn than the same labor expended on tired ground. So planters cleared their forested acreage as a way of keeping the productivity of slaves as high as possible, which is why plantations needed to be so much larger than their cleared acres (the ones in production at any given time). The problem was that few planters possessed enough acreage to shift an arable parcel as large as 300 acres through their wooded land in a fifteen- or twenty-year cycle (a reasonable period for the forest to return before being burned and the soil cultivated again) without running out of land. When they let this balance slip, planters packed their households and moved to Alabama, Mississippi, and Texas—the planters' frontier. This is the formula for southern migration, and it was driven by the collision between slavery and forested land.[17]

Abandoned cotton fields reverted to forest, and this seemed to threaten the old distinction between settled space and wilderness. The possibility that pines and hardwoods might reclaim lands first cleared of them a century or more before evoked an emotional response. Southern planters with no plans to strike out for the West cried out that migration had caused a backward shift in the expected course of civilization, understood since the eighteenth-century Scottish philosophers as a forward march in stages from primitive to sophisticated forms of economy. Wrote the Agricultural Society of South Carolina, "A natural fertility [in the soil] may soon be impaired by injudicious treatment . . . complete exhaustion will at length compel to a total abandonment. It is easy to foresee that an extensive practice of this kind will, in the course of a few years, greatly impoverish a country, and finally convert it almost into a wilderness."[18] Human history, noted the planters, presented other examples of the same dismal process—the deserts of North Africa and the plains of Italy had once been fertile. "We have too many similar examples of this melancholy truth, when we behold, in our own country, extensive tracts of land covered with broom-grass, and abandoned." Another author suggested that the waste of southern agriculture had transported the poor white farmer back to an earlier stage of social evolution by reducing him to "a hunter upon the hills."[19]

Forests emerge in the literature of improvement as forces of balance in the landscape, as controls for the retention of moisture and stores of wood, and yet also as harbingers of social decline whenever they escaped the narrow position defined for them. Forests served agrarian society; they should never arise atop its failure, as some planters believed had happened throughout the Atlantic South. As for the western wilderness—the undomesticated forests and prairies that lay over the Appalachian Mountains and the Blue Ridge—farmers feared it as a population vacuum and certainly wondered about it, but otherwise they gave it little thought. The important thing to note here is that the ethic of conservation taking shape among eastern farmers did not have to do with trees or even with soils as such but with society as its members understood it and as it benefited them. Yet within their desire to maintain a density of population befitting a long-established country was born a larger vision of permanence—a *Landschaft*—that eventually included wilderness.

The seminal thinker in what became conservation was a classical scholar and member of Congress from Vermont named George Perkins Marsh. He had no love for the backwoods and displayed no interest in wilderness as a refuge or a moral counterweight to industrial society. Marsh hated everything about the West, declared it a barren waste, loathed it as a drain on the energies of government, and behaved as though it were a threat to the political and cultural power of New England. His only journey to the prairie, in 1837, came to a halt at the Falls of St. Anthony (now St. Paul), where he declared the West uninhabitable. From out of this fear and loathing came an idea. If Marsh could wave his hands and prevent migration out of New England, he would have to find a way for the people there to maintain their supply of trees. Marsh came to know what happened when a hillside lost its forest cover: its soils came sliding down with the rains into streams and rivers, never to return. So, just like the improving farmers and planters, Marsh set out to conserve his society, tied to the particular resources that would allow it to endure.

Marsh redefined the purpose of forests. In an address delivered to his constituents during the Mexican War, he made the point that leaving trees standing caused a more sustained progress than cutting them down: "The increasing value of timber and fuel ought to teach us, that trees are no longer what they were in our fathers' time, an encumbrance." Not only that, but trees literally held the landscape together: "Steep hill-sides and rocky ledges are well suited to the permanent growth of wood, but when . . . they are improvidently stripped of this protection, the action of sun and wind and rain soon deprives them of their thin coating of vegetable mould, and this, when exhausted, cannot be restored by ordinary husbandry."[20] He believed that migrants abandoned more than their homes by their use-it-and-leave-it notion of private wealth; they abandoned a larger environment for which they were responsible.

Yet though he first articulated a formal philosophy of conservation, Marsh was not the first to suggest that farmland and woodland might coexist. A number of antebellum farmers, besides Jesse Buel, expressed similar ideas. Perhaps the most impressive was a Pennsylvania farmer named John Lorain whose book, *Nature and Reason Harmonized in the Practice of Husbandry* (1825), defined an American variation on the sophisticated methods of English agriculture. Lorain wrote about timber and forests like a modern-day agent of the Department of Agriculture:

> In new settlements, the timber is commonly the greatest obstacle to cultivation; this begets an emulation to destroy it; he is considered the best farmer, who clears the most land. This enterprise would be laudable to a

> certain extent; the habit, however, of considering the timber in the way, induces the farmer to wage a perpetual war against it, until his eyes are opened, by finding that he has neither fencing nor firewood left.[21]

Leaving the trees standing would enhance the farmer's material well-being, and though Lorain had little of Marsh's insight into hydrology, he understood the modern formula for balance in the landscape. Intensive cultivation, represented in the 1820s by manure piles and first-rate pasture grasses, saved trees and provided better for the farmer by making further clearing or migration unnecessary. Lorain had seen firsthand the process he most feared. He lived in western Pennsylvania, closer to Pittsburgh than to Philadelphia, where he watched backwoods people. The experience brought him to describe local behavior in universal terms: "Man is the most destructive animal in the universe, when he considers that his resources cannot fail."[22]

Yet it was Marsh who first scrutinized the relationship between agriculture and woodland. He deepened his thought over the next twenty years, culminating in the publication of *Man and Nature* (1864), one of the great works of American scholarship in the nineteenth century. *Man and Nature* makes the argument that humans had changed the world on a geological scale, in ways that could never be reversed. Like the older Buel, Marsh derived his ideas from Whig principles stressing responsibility not just for one's own affairs and fortune but also for the greater society. Conservation simply extended these concerns from education and the political culture to forests and soils.

When the forest reclaimed farmland throughout New England, Marsh refused to read its return as a sign that civilization had failed its mission. Competition in the grain market between New England and Ohio, made decisive by New England's many disadvantages for commercial grain cultivation, forced the conversion of land to more intensive uses or its reforestation beginning in the 1850s and intensifying after the Civil War. He repeated what many northern farmers had said since the 1820s—that too much land had been cleared in the first place, much of it marginal for agriculture. Its regrowth could only be a benefit:

> When we consider the immense collateral advantages derived from the presence [of the forest], the terrible evils necessarily resulting from the destruction of the forest, both the preservation of existing woods, and the far more costly extension of them . . . are among the most obvious of the duties which this age owes to those that are to come after it.

In an echo of Buel and Lorain, who saw a greater landscape of which their farms formed a part, Marsh saw unshorn forests as the basis of an enduring civilization.[23] The conservationist thought of the eastern farmers suggested the integrated approach to land use that would become dominant among conservationists in the next century.

Wilderness, Agriculture, and Romantic Art

Conservationist thinkers and antebellum farmers were not the only people to think about the collision between agrarian society and the American forest in the nineteenth century. Romantic landscape painters articulated their own conception of agriculture, relating it to national and human destiny. Fearful of the pace and direction of industrial society, romantics made wilderness the repository of values that people once invested in the monarchy or the church. American painters identified the forest with moral purity and social stability.

Wilderness romanticism only appeared to be an all-or-nothing proposition. In fact, the first romantic thinkers found inspiration in the English countryside, a thoroughly domesticated place. Yet Thomas Cole asserted a severe counternarrative to the dominant thinking about progress and civil society popular during the 1830s and 1840s (see chapter 6). By the time he began to paint American landscapes, Cole had absorbed the insights of the Scottish moralists and their belief that all societies pass through material stages—from savagery to civilization—including the invention of agriculture. Cole turned the Scots on their heads, arguing that because agriculture brought about the culmination of progress—whether understood as industrial capitalism or political empire—it stood for a kind of original sin. It introduced accumulation and wealth and eventually led to the forest-destroying, morally corrupted, and self-destructive state.

Romantic painters repeatedly turned to the countryside for the meaning it contained about the American future. The moment when farmers arrived in the wilderness to clear it and establish their homesteads became a crucial moment for philosophical musings, and Cole, in particular, responded with highly symbolic canvases. Notable among his paintings depicting civilization's opening in the forest is *Home in the Woods* (1847). An isolated family lives at the edge of a lake. The father comes home with his fishing pole and greets the mother and three children, who wait for him at the door of their log cabin. The family owns no animals, plants no crops, and clears no trees from the lakeside. Contradicting every assumption that a people must practice agriculture on their way to civilization, Cole strongly suggests that a wilderness existence based on hunting and fishing maintained the family and fostered stability and sufficiency.

Other painters shared none of Cole's despair and painted economic activity as existing within the epic scale of nature. In Frederic Church's *Haying Near*

New Haven (1849), people gathering hay in a river meadow appear tiny under the prow of West Rock. Their activities feel almost incidental. The forest does not recoil, but neither does it embrace them. The mood is blank. Other painters went so far as to admit industrialism into the frame, often in the form of railroads and viaducts, suggesting that these had become "natural" elements. Even in celebration, though, romantic painters expressed a fascination with farmland as the navel of modernity, as a practice that seemed to exist between the forces of creation and destruction.

Conclusion

The opening in the woods that came to shore with the first Europeans needed fresh land to feed it. Agrarian people continually sought territory in order to replicate their patterns and institutions for the generations that would make their homes over the mountains. Their own increase, even more than their depletion of soil nutrients, impelled them outward.[24] And because this outward "errand in the wilderness" carried such moral weight from the start, it took on symbolic meaning. The farmer became an icon of democracy, and the agrarian dream he signified interpreted the cutting of the forest as an act of cultural creation, as a matching of light for light with God, carried on by a people for whom dark bowers represented a veil separating them from grace. By providing the tools and weapons for this moral and ecological reclamation, agriculture did its age-old work.

We have read this conflict into American history, from Thomas Jefferson's motives for the Louisiana Purchase, to the first conservationist thinkers among eastern farmers of the 1820s, to the romantic paintings that interpreted agriculture as an incursion into the social and ecological stability of primitivism. Agrarian settlement continued to make demands on territory. Irrigated agriculture claimed the riparian ecosystems of the Central Valley by the 1890s, followed by the desert lands from the Mexican border to central Oregon and Idaho by early in the next century. The sod-house dwellers on the Great Plains came in large numbers during the 1870s, as the government hunted bison and sequestered Indians to make room for them.

At the same time, a different point of view began to make an imprint. Stemming from Marsh's embrace of forests as crucial to the whole landscape came a number of other developments. Congress created the first national park in 1872 from a region cut out of the northwest corner of the state of Wyoming. Yellowstone National Park (along with other early parks, like Yosemite in California and the Adirondacks in New York) represented the first time that Congress imposed legal limits on the spread of agricultural settlement. The park created an entirely new category of land use—protected wilderness—where no crops or domestic animals would be allowed. The government explorer and anthropologist John Wesley Powell argued in a report of 1879 that "it may be doubtful whether, on the whole, agriculture will prove remunerative"[25] in the arid regions, where rainfall was below twenty inches a year. The park imposed legal limits, but Powell said that ecological limits existed to the colonization of the West by farmers.

By the beginning of the twentieth century, a number of observers attempted to recast the old blood-and-guts tendencies of migrating farmers from conquest to coexistence. The romantic agronomy of Liberty Hyde Bailey; the genteel but prescient *Garden and Forest* magazine (published beginning in 1888 by Charles S. Sargent of the Arnold Arboretum); and the regional planning of Lewis Mumford expressed visions of prosperity without exploitation. Mumford virtually invented what we today call *human ecology*. As he wrote in 1925, "Regional planning is the New Conservation—the conservation of human values hand in hand with natural resources. . . . Permanent agriculture instead of land skinning, permanent forestry instead of timber mining, permanent human communities."[26]

There is no better example of the new détente than the Harvard Forest, founded in 1907 by Richard T. Fisher. In the 1920s, Fisher and his colleagues concluded that the landscape of central Massachusetts told a complicated story, one that had as much to do with human as natural history. In order to understand the forest, they needed to see it unfold, so they constructed a series of visual models narrating one location over two centuries. The dioramas are each somewhat smaller than a department store window and feature remarkable detail, down to the human figures performing all sorts of activities. The story begins with the ancient forest (1700), followed by the first English settlement (1720s), then farm abandonment (1850s), followed by the return of old-field white pines (1910), the growth of hardwoods after the pines had been harvested for box boards (1915), and finally a maturing hardwood forest (1930). Amid the violence between farm and forest, this all-encompassing conception of landscape dynamics counts as a true innovation. By embracing all forms of land use and the return of the forest to New England, they pointed toward a conception of ecology that included human societies and a conception of agriculture that included forests—an armistice, if not a durable peace.[27]

Notes

1. Robert Pogue Harrison, *Forests: The Shadow of Civilization* (Chicago: University of Chicago Press, 1992); Hugh Brody, *The Other Side of Eden: Hunters, Farmers, and the Shaping of the World* (Vancouver: Douglas & McIntyre, 2000).

2. See Brian Donahue, *The Great Meadow: Farmers and the Land in Colonial Concord* (New Haven, Conn.: Yale University Press, 2004).

3. On animals, see Virginia DeJohn Anderson, *Creatures of Empire: How Domestic Animals Transformed Early America* (New York: Oxford University Press, 2004); and Elinor G. K. Melville, *A Plague of Sheep: Environmental Consequences of the Conquest of Mexico* (New York: Cambridge University Press, 1994). Also see William Cronon, *Changes in the Land: Indians, Colonists, and the Ecology of New England* (New York: Hill and Wang, 1983). On population and intensive cultivation, see Ester Boserup, *The Conditions of Agricultural Growth: The Economics of Agrarian Change under Population Pressure* (New York: Aldine, 1965).

4. John F. Richards, *The Unending Frontier: An Environmental History of the Early Modern World* (Berkeley and Los Angeles: University of California Press, 2003).

5. Thomas Jefferson, *Notes on the State of Virginia* (1782), Electronic Text Center, University of Virginia Library (http://etext.virginia.edu/toc/modeng/public/JefVirg.html), Query 19.

6. Roger G. Kennedy, *Mr. Jefferson's Lost Cause: Land, Farmers, Slavery, and the Louisiana Purchase* (New York: Oxford University Press, 2003), 41–42.

7. I do not know that Jefferson saw the fires of migrants burning on the Blue Ridge, but I do know that others saw them. See Steven Stoll, *Larding the Lean Earth: Soil and Society in Nineteenth-Century America* (New York: Hill and Wang, 2002), 145.

8. Melville, *A Plague of Sheep*; Thomas R. Hietala, *Manifest Design: Anxious Aggrandizement in Late Jacksonian America* (Ithaca, N.Y.: Cornell University Press, 1985).

9. On shifting cultivation, see Harold C. Conklin, *Hanunóo Agriculture: A Report on an Integral System of Shifting Cultivation in the Philippines* (Rome: Food and Agriculture Organization of the United Nations, 1957; repr., Northford, Conn.: Elliot's Books, 1975). For American fire practices, see Stephen Pyne, *Fire in America: A Cultural History of Woodland and Rural Fire* (Princeton, N.J.: Princeton University Press, 1982).

10. Jefferson wrote the Ordinance of 1784 (which established the territorial process, through which the territories could become states) and the Ordinance of 1785 (which established the survey system) but not the Ordinance of 1787 (which elaborated on the earlier legislation and applied directly to the lands north of the Ohio River). On the politics of slavery surrounding both ordinances, see Kennedy, *Mr. Jefferson's Lost Cause*, 249. Also see Kennedy's sources, including T. C. Pease, "The Ordinance of 1787," *Mississippi Valley Historical Review* 25 (1938): 167.

11. J. A. Etzler, *The Paradise within the Reach of All Men, without Labor, by Powers of Nature and Machinery* (Pittsburgh: Etzler and Reinhold, 1833), 1.

12. Henry David Thoreau thought enough of Etzler to review his work in the *United States Magazine and Democratic Review* (November 1843), 451. Etzler helped to found the Tropical Emigration Society and moved a company of English emigrants to Venezuela in 1845. When the colony broke up the following year, he disappeared. See *The Collected Works of John Adolphus Etzler: Facsimile Reproductions with an Introduction by Joel Nydahl* (New York: Scholars' Facsimiles and Reprints, 1977).

13. The clearing of so much wood so quickly might have changed more than the landscape. Global temperatures spiked in 1850 after four hundred years of lower-than-

average temperatures, and they have not fallen since. The reason cannot be attributed to coal-burning factories, which accounted for a minuscule amount of carbon emissions until the end of the nineteenth century. Instead, it might be owed to backwoods families and other settler cultures worldwide, which cut and burned millions of acres, releasing one of the world's largest stores of carbon into the atmosphere at an unprecedented rate. The year 1850 is generally regarded by geologists as the last year of the Little Ice Age, an epoch that began in 1350 and was characterized by periods of flood, famine, and July snowstorms throughout the Northern Hemisphere. Atmospheric scientists now believe that some of the warming since 1850 is due to the end of the Little Ice Age, but not all of it. See Vaclav Smil, *Cycles of Life: Civilization and the Biosphere* (New York: Scientific American Library, 1997), 85; Brian Fagen, *The Little Ice Age: How Climate Made History* (New York: Basic, 2001). And yet this does not mean that swidden cultivators should be blamed for global warming. The growth of young forest absorbs more carbon dioxide than mature stands, suggesting that swidden farming might help to sequester more carbon than it releases to the atmosphere. Settlement cultures, like those in North America, tended not to engage in a true swidden cycle, in which the forest grew back before being reduced to ashes again. Rather, they cleared land once and for all.

14. See Stoll, *Larding the Lean Earth.*

15. *American Farmer,* 2d ser., 1 (1839).

16. Jesse Buel, *The Farmer's Companion; or, Essays on the Principle and Practice of American Husbandry,* 6th ed. (New York: Harper, 1847), 55.

17. Highland farmers who had no slaves also migrated for land, though the exact mechanism was different. Very often, they found themselves with poor or highly eroded Piedmont soils.

18. *Papers Published by Order of the Agricultural Society of South Carolina* (Columbia, S.C.: Telescope, 1818), 6.

19. *Memoirs of the Society of Virginia for Promoting Agriculture Containing Communications on Various Subjects in Husbandry and Rural Affairs* (Richmond, Va.: Shepherd and Pollard, 1818). See George M. Weston, *The Poor Whites of the South* (Washington, D.C.: Buell and Blanchard, 1856).

20. George P. Marsh, *Address Delivered before the Agricultural Society of Rutland County, September 30, 1847* (Rutland, Vt., 1848), 17.

21. John Lorain, *Nature and Reason Harmonized in the Practice of Husbandry* (Philadelphia: Carey and Lea, 1825), 333.

22. Lorain, *Nature and Reason Harmonized in the Practice of Husbandry,* 333.

23. George Perkins Marsh, *Man and Nature* (Cambridge, Mass.: Harvard University Press, 1965), 279–80.

24. Donahue, "A Town of Limits," in *The Great Meadow.* On the colonizing tendency behind American agriculture of the nineteenth century, with special reference to the U.S. Department of Agriculture, see Frieda Knobloch, *The Culture of Wilderness: Agriculture as Colonization in the American West* (Chapel Hill: University of North Carolina Press, 1996).

25. John Wesley Powell, *Lands of the Arid Region of the United States* (Washington, D.C., 1879), 3.

Five

Natural History, Romanticism, and Thoreau

Bradley P. Dean

Twenty-three decades after the Pilgrims arrived in Provincetown harbor to a "hideous and desolate wilderness," another sort of pilgrim returned to his study in that same New World. He recorded a different impression of what he had seen during his afternoon walk: "How near to good is what is wild. There is the marrow of nature—there her divine liquors—that is the wine I love." What had been "hideous and desolate" in 1620 had in 1849 become "near to good."[1]

Wilderness is a manifestation of the physical world that people *perceive* very similarly, but each person's *conceptions* of it differ. Wilderness per se does not change; we change. The newly arrived Pilgrims felt threatened because of conceptions about wilderness that had been packed in their minds just as surely as food and other necessities had been packed in the hold of their ship. Humans see in wilderness what we bring to it in our minds. And in the years since the Pilgrims landed in the New World, new conceptions of wilderness based upon changing religious, scientific, and philosophical beliefs about nature—in particular, natural history and romanticism—had led some mid-nineteenth-century Americans to a radically different way of seeing wilderness.

As Old World offshoots of the Protestant Reformation transplanted in the New World, the Pilgrims conceptualized wilderness as a threateningly mysterious domain to be conquered by religion and European-style agriculture. During the seventeenth century, as Americans transformed vast tracts of wilderness into farmscapes surrounding small church-centered towns, their religiocentric world view was gradually supplanted by an emerging faith in the possibilities of human reason, a faith stemming primarily from the achievements of early natural historians (now called scientists) and the impressive practical results of those achievements. America's first learned organization, the American Philosophical Society, exemplified this faith in its mission to "promote useful knowledge" by pursuing "all philosophical [i.e., scientific] Experiments that let Light into the

Nature of Things, tend to increase the Power of Man over Matter, and multiply the Conveniencies or Pleasures of Life." This Enlightenment era world view, which conceptualized wilderness as a benignly comprehensible domain to be exploited by reason and technology, was coupled with deism, a religious perspective which posited that God had made the universe and set it working on the basis of scientifically discernible laws. Because these laws, like nature itself, originated in and reflected the mind of God, the study of nature took on tremendous religious significance, as suggested by Mark Stoll in chapter 3 and as seen in the title of John Ray's influential treatise *The Wisdom of God Manifested in the Works of the Creation* (1691). Cotton Mather, Ray's foremost proponent in America, argued in 1720 that nature study complements Christian Scriptures, for they are "the *Twofold Book* of GOD." Since nature showcases God's wisdom, wilderness—nature unadulterated by humanity—is where divine wisdom is best studied.[2]

Explicitly or implicitly, Enlightenment era naturalists subscribed to natural theology, the study of nature for evidence of divine design. Their first task, inventorying the constituent parts of nature, was greatly facilitated by the development and widespread acceptance of Carl Linnaeus's hierarchical, binomial classification system, still in use (with many modifications) by scientists today. Linnaeus, himself an ardent natural theologian, inspired generations of specimen collectors and arranged for nineteen of his own students to travel the globe, often accompanying government-sponsored voyages of discovery or trade, but always sending specimens to their mentor in Sweden. One of these students, Pehr Kalm, collected specimens in America from 1748 to 1751 and published the earliest account of American wilderness written by a professionally trained natural historian.[3]

Well before Kalm's arrival, however, pre-Linnaean natural historians were combing the New World's wilderness and cataloging its bounties, although not according to the Linnaean system. Preeminent among these early naturalists was American-born John Bartram, who in 1728 planted his celebrated botanical garden south of Philadelphia and in 1732 began sending specimens to wealthy European patrons, eventually supplying more than fifty "subscribers," including Linnaeus himself, who hailed Bartram as "the greatest natural botanist in the world." In 1751, Bartram published an account of his 1743 excursion to Lake Ontario, with Kalm contributing an appendix based on his 1750 visit to Niagara. Almost exclusively factual, *Observations on the Inhabitants* presaged more-literary treatments of American wilderness that followed from the pens of natural historians, such as Bartram's son William, during the coming decades.[4]

During the forty years between John Bartram's *Observations on the Inhabitants* and William Bartram's own *Travels through North and South Carolina* (1791), another gradual change occurred in Western conceptions of wilderness. Among the many factors contributing to that change, two are especially important. First, the early romantic Edmund Burke provided naturalists with a physiological explanation for the sensations they experienced when viewing wilderness. Generally, Burke asserted that nature's sublime features cause a delightful tightening of the body's fibers, but nature's beautiful features pleasingly relax those fibers—an explanation which validated aesthetic responses to wilderness. Second, whereas Reformation era Americans transformed wilderness into bucolic landscapes, Enlightenment era Americans developed technologies that resulted in increasingly industrialized landscapes. Although pastoral landscapes tend not to alienate humans, industrialized landscapes generally do, and when humans feel physically safe but psychically alienated, they seek palliatives to ameliorate their sense of alienation.[5]

Bartram's *Travels through North and South Carolina*, a literary, scientific redaction of a four-year journey through America's southeastern frontier (1773–1777), reflects the larger cultural transition from an Enlightenment to a romantic world view better than the work of any other American naturalist because he articulates conceptions of wilderness in a hybrid fashion: as a natural theologian with romantic sensibilities. His assumptions about nature are evident in his introduction: "nature is the work of God omnipotent; and . . . even this world, is comparatively but a very minute part of his works. If then . . . the mere material part, is so admirably beautiful, harmonious, and incomprehensible, what must be the intellectual system? . . . this must be divine and immortal[.]" Bartram as a man of the Enlightenment is seen throughout *Travels* as he reflects upon the benign order and harmonies of the universe, one aspect of which is the inevitable transformation of large tracts of wilderness into "the most populous and delightful seats on earth." His optimistic faith in reason is pervasive, but his assertion that "the mere material part" of the world is "incomprehensible" marks him as a nascent romantic, as do the highly subjective, often Burkean responses to natural phenomena evident throughout his book.[6]

Although naturalists after Bartram described American wilderness in many genres, by far the bulk of their writings took the form of exploration or travel narratives like those of the Bartrams. A significant number of these early nineteenth-century narratives were reports commissioned by the government, but far more were penned by independent naturalists capitalizing on a reading public whose appetite for narratives of wilderness exploration seemed insatiable. Yet these narratives were only one of many manifestations of the cult of nature that developed in the early nineteenth century. Landscape painting surged in popularity during these years, as did the imaginative works of poets and fiction writers, such as the nature poems of William Cullen Bryant and the frontier novels of James Fenimore Cooper. Perhaps the most dramatic manifestation of the heightening interest in nature during this period was the springing up of the tourism industry in antebellum America, with railroads shuttling city dwellers to and from wilderness and seaside resorts. To take one example, in 1836, the White Mountains region of New Hampshire featured one small inn, but by the mid-1850s six lavishly appointed "mountain houses" accommodated trainloads of tourists from the country's burgeoning cities. Nature, and particularly wilderness, had become the Romantic Era's palliative. Americans of that era conceptualized wilderness as a mysteriously redemptive domain to be enjoyed and celebrated for its ability to assuage the dehumanizing excesses of their increasingly urban, increasingly industrial civilization.[7]

During the 1820s, authors and artists began turning to the new country's wilds as a potential source of its cultural greatness. In 1823, America's first renowned painter, Thomas Cole, began his landscapes, and that same year Cooper published *The Pioneers*, the first of his enormously popular Leatherstocking Tales and the book that established him as the country's first bestselling novelist. Despite the impressive achievements of Cole and Cooper, however, America's cultural independence was not achieved until shortly after the Boston-born writer and philosopher Ralph Waldo Emerson moved to Concord in 1834. His seminal book *Nature* (1836), widely regarded as the first literary and philosophically sophisticated articulation of distinctively American ideas, inaugurated what has since become known as the American Renaissance, which lasted until the Civil War and saw the publication of such classics of American literature as Hawthorne's *The Scarlet Letter* (1850), Melville's *Moby-Dick* (1851), Thoreau's *Walden* (1854), and Whitman's *Leaves of Grass* (1855).

Emerson's principal protégé was a native of Concord, having been born there in 1817. He graduated from Harvard University in 1837 and was known locally as an inventive engineer, an uncannily accurate surveyor, and an astute naturalist who wrote two books: one a commercial disaster, the other more successful but decidedly unusual, in part because he portrayed himself as having lived for two years (1845–1847) as a hermit in the woods beside a pond just a mile and a half south of Concord Center. For decades after his death in 1862, Henry David Thoreau was regarded as a minor author, a pale imitator of Emerson. But his reputation grew considerably around the turn of the twentieth century, after his complete writings were first published in 1906. Since the 1960s, he has come to be regarded not just as a preeminent American writer but also as a key figure in the history of American wilderness. His conceptions of wilderness are complex, and we can best begin to grasp them by understanding how his mind was conditioned to interpret what he saw when he first significantly encountered wilderness.

Thoreau came by his love of nature honestly. Both of his parents reportedly "had a common interest in nature" and early in their marriage "could often be found in their spare time, at almost any season of the year," exploring the countryside around Concord. They imparted their interest in nature to their young son, who later testified eloquently to his early attachment to the natural environment: "Formerly methought nature developed as I developed and grew up with me. My life was ecstasy. In youth, before I lost any of my senses, I can remember that I was all alive and inhabited my body with inexpressible satisfaction; both its weariness and its refreshment were sweet to me." Thoreau's attraction to nature was somewhat unusual because he was particularly attracted to wild settings rather than pastoral ones. "It does seem as if mine were a peculiarly wild nature, which so yearns toward all wildness," he wrote in December 1841. The distinctive personality that resulted from these yearnings was noticed by friends and acquaintances well before Thoreau's Walden years. Nathaniel Hawthorne referred in April 1843 to his young friend's "wild freedom," for example, and found Thoreau "one of the few persons . . . with whom to hold intercourse is like hearing the wind among the boughs of a forest-tree."[8]

Although Thoreau was raised with an interest in nature and as a youth evinced a temperamental attraction to nature's wilder settings, his most important conceptions of nature were not formed until his final year of college. On April 3, 1837, he withdrew Emerson's *Nature* from one of Harvard's libraries. This diminutive volume eloquently and concisely sets forth the basic principles of American transcendentalism. The book revolutionized Thoreau's intellectual life and provided him with a worldview that he expanded in distinctive ways but never forsook. To understand how he earned his status as America's first apostle of wilderness, it is necessary to understand a few fundamentals of Emersonian transcendentalism.

Emerson begins *Nature* by challenging his readers to jettison their inherited perspectives on foundational matters such as poetry, philosophy, and religion in favor of fresh, new perspectives, which he encouraged his readers to develop themselves. The goal was to develop "an original relation to the universe," which Emerson and other transcendentalists, including Thoreau, regarded as a moral and intellectual imperative. Emerson defined the universe as consisting of matter and spirit. Matter included his body, other people, and nature—what Emerson called the "NOT ME." Wilderness and civilization, or nature and art, were matter as well. Emerson defined *wilderness* as "essences unchanged by man," and *civilization* as "the mixture of [man's] will with" those "essences."[9]

Two implications necessarily follow from Emerson's definitions. First, each individual human being is a spirit existing within that portion of "the NOT ME" called "my own body." Later, in the "Spirit" section of *Nature*, Emerson specifically refers to the body as a dwelling: "As we degenerate [physically and morally], the contrast between us and our house [our body] is more evident." And we saw that Thoreau remembered his youth as a time when "I was all alive and *inhabited my body* with inexpressible satisfaction" (emphasis added). This conception of human identity was much more than a metaphor to both men; it is the basic human condition—*who* we are (spirits) and *where* we are (in bodies). Because it also follows from Emerson's definitions that we as spirits relate to the physical universe through the agency of our bodies, the opening challenge of *Nature* can be recast in reference to wilderness as follows: how can we as spirits through the agency of our bodies develop "an original relation" to those portions of nature whose "essences" are more or less "unchanged by man"?[10] Thoreau's lifelong effort to answer this question accounts for his status as the patron saint of wilderness.

Thoreau's first significant encounter with wilderness occurred on September 8, 1846, when he climbed to the top of a ridge near the summit of Mount Katahdin in Maine. Nature there was unlike anything he had encountered before. "It was vast, Titanic, and such as man never inhabits," he wrote. "The tops of mountains are among the unfinished parts of the globe, whither it is a slight insult to the gods to climb and pry into their secrets"—an insult in part because the initial processes of creation seemed to be taking place there. Nature on mountaintops "seems to say sternly, why came ye here before your time? This ground is not prepared for you." The extreme wilderness atop Katahdin is a sanctuary where mysterious, creative powers are engaged in the earliest stages of transmuting pure, elemental matter into the raw materials upon which human life depends.[11]

During his descent, Thoreau crossed a section of "exceedingly wild and desolate" burnt lands and suggested that perhaps then, rather than atop the ridge, he "most fully realized that this was primeval, untamed, and forever untameable *Nature*, or whatever else men call it." He found himself traversing the burnt lands "familiarly, like some pasture run to waste, or partially reclaimed by man"; but then he reflected that no *human* made or claimed this land: "Here was no man's garden, but the unhandselled globe. . . . It was the fresh and natural surface of the planet Earth, as it was made for ever and ever,—to be the dwelling of man, we say,—so Nature made it, and man may use it if he can." Wrapping up his description with the observation that what he saw there on the slope of Katahdin "was a specimen of what God saw fit to make this world," he moved from description to an interpretation of what raw wilderness *means*:

> What is it to be admitted to a museum, to see a myriad of particular things, compared with being shown some star's surface, some hard matter in its home! I stand in awe of my body, this matter to which I am bound has become so strange to me. I fear not spirits, ghosts, of which I am one,—*that* my body might,—but I fear bodies, I tremble to meet them. What is this Titan that has possession of me? Talk of mysteries!—Think of our life in nature,—daily to be shown matter, to come in contact with it,—rocks, trees,

wind on our cheeks! the *solid* earth! the *actual* world! the *common sense! Contact! Contact! Who* are we? *where* are we?[12]

This "Contact" passage has stimulated a great deal of controversy among scholars, with some commentators suggesting that Thoreau had been so traumatized by his contact with raw wilderness that he was unable to control even his syntax. But the passage is no extemporaneous effusion. Instead, it is a carefully crafted text written by an extraordinarily talented Emersonian transcendentalist intent on conveying to his readers a sense of the original, wilderness-inspired relation to the universe that he experienced on the mountain. In November 1857, more than eleven years after this experience and just four and a half years before his death, Thoreau pointed out in a letter to a friend who had recently climbed Mount Washington in New Hampshire, "It is after we get home that we really go over the mountain, if ever. What did the mountain say? What did the mountain do?" He clearly had his Katahdin experience in mind, for in the preceding paragraph, after pointing out that he felt "the same awe when on [mountain] summits that many do on entering a church," he wrote, significantly, "To see what kind of earth that is on which you have a house and garden somewhere, perchance! . . . You must ascend a mountain to learn your relation to matter, and so to your own body, for *it* is at home there, though *you* are not." The Emersonian legacy here is obvious. Clearly, Thoreau's contact with wilderness on Katahdin compelled him to realize who and where he was, clarifying his sense of himself as a spirit inhabiting a body and making him realize very acutely that while his body was at home in "the *actual* world" of matter, his spirit was emphatically *not* at home in such "unfinished parts of the globe." His 1857 letter indicates that his wilderness encounter of 1846 had been intensely redemptive, not alienating, and that the encounter continued to inform his thinking. "I go up there [to mountaintops] to see my body's cousins," he told his friend. "There are some fingers, toes, bowels, etc., that I take an interest in, and therefore I am interested in all their relations."[13]

This notion of "relations" is critical. In "The Village" chapter of *Walden*, he explained that getting physically lost can be a salutary experience because "[n]ot till we are lost, in other words, not till we have lost the world [defamiliarized the relations between our spirits and the world], do we begin to find ourselves [learn we are spirits], and realize where we are [dwelling within bodies] and the infinite extent of our relations" with the rest of the universe. The miracle of human existence to which Thoreau calls attention is not that we are spirits, but that we as spiritual beings have contact with or relations to the world of matter. A person's body is the locus of these relations, the agency by which spirit contacts matter. And because the universe is infinite, the extent of those relations is likewise infinite—and thereby miraculous. This insight provided the basis for the "original relation to the universe" that Thoreau went on to articulate with increasing sophistication for the remainder of his life.[14]

Thoreau's ascent of Katahdin took place halfway through his famous twenty-six-month sojourn at Walden Pond. After returning from Maine to his house on the shore of Walden, he wrote the account of his experience on the mountain—and added a highly significant paragraph to the account he was writing of his experiment at the pond. "Our village life would stagnate if it were not for the unexplored forests and meadows which surround it," that paragraph begins. "We need the tonic of wildness. . . . We must be refreshed by the sight of inexhaustible vigor, vast and Titanic features, the sea-coast with its wrecks, the wilderness with its living and its decaying trees, the thunder cloud, and the rain which lasts three weeks and produces freshets." His account of Maine's "grim, untrodden wilderness" as a "tangled labyrinth of living, fallen, and decaying trees" indicates clearly that he wrote the *Walden* paragraph with his recent trip in mind. And in this tonic-of-wildness paragraph, he articulates another lesson that he learned from his contact with wilderness in Maine. As the paragraph proceeds, he speaks of being "cheered" not just by death, but death on a grand, Malthusian scale: "I love to see that Nature is so rife with life that myriads can be afforded to be sacrificed and suffered to prey on one another; that tender organizations can be so serenely squashed out of existence like pulp,—tadpoles which herons gobble up, and tortoises and toads run over in the road; and that sometimes it has rained flesh and blood!" Just as raw matter atop mountains teaches the lesson of our spiritual identity, the prevalence of death in the universe teaches the lesson of life. Trees live and die, and their decay contributes to a natural cycle. Transmuting death into life—the cycle of life, death, and rebirth—reflects nature's "strong appetite and inviolable health." What, then, shall we make of death? "The impression made on a wise man," Thoreau states flatly, "is that of universal innocence." What might be called his eco-transcendentalist world view does not invalidate Malthus's economic calculations; it simply expands the Malthusian equation to encompass a view of nature as an endless, symbiotic cycling and recycling of matter, a Thoreauvian eternal return.[15]

On his way to Cape Cod in October 1849, two years after leaving Walden, Thoreau witnessed the aftermath of a shipwreck at Cohasset. He described a landscape strewn with human corpses placed in makeshift coffins, just as he had described Katahdin's vast, terrific, rock-strewn landscape. But whereas he had carefully crafted his mountainside description to be allusive and impassioned, he restricted his description of this seaside disaster to dispassionate reportage: "I saw many marble feet and matted heads as the cloths were raised, and one livid, swollen, and mangled body of a drowned girl . . . the coiled-up wreck of a human hulk, gashed by the rocks or fishes, so that the bone and muscle were exposed, but quite bloodless,—merely red and white,—with wide-open and staring eyes, yet lustreless, dead-lights." And then, just as he had with the mountain, he began to interpret what he saw: "On the whole, it was not so impressive a scene as I might have expected. . . . If this was the law of Nature, why waste any time in awe or pity?" The interpretive passages evince no indication whatever of an emotional struggle about the horrible brutality of a ferocious agency killing innocent seafarers. Instead, Thoreau unflinchingly conveys his interpretation of the prevalence of death—and in this instance, the deaths of humans, not trees:

> Why care for these dead bodies? They really have no friends but the worms or fishes. Their owners were coming to the New World, as Columbus and the Pilgrims did,—they were within a mile of its shores; but, before they could reach it, they emigrated to a newer world than ever Columbus dreamed of, yet one of whose existence we believe that there is far more universal and convincing evidence—though it has not yet been discovered by science—than Columbus had of this.

In these introductory passages to *Cape Cod*, there is undeniably a looping out of the eternal return: spirits depart for "a newer world"; their corpses remain behind to participate in the natural cycle by nourishing "worms or fishes." The implication is that the eternal return applies to the physical world only: our bodies participate in it, but our spirits are released to *another* world. "It is hard to part with one's body," he concludes, "but, no doubt, it is easy enough to do without it when once it is gone. . . . The strongest wind cannot stagger a Spirit; it is a Spirit's breath."[16]

The view that a physical fact, such as the wind, corresponds to or suggests a spiritual corollary reflects Thoreau's Emersonian legacy. And it is precisely here that his distinctive contribution to a history of wilderness begins to become apparent. Emerson provided him a congenial theory of the universe but an uncongenial methodology for learning how the universe works. Emerson was greatly enamored of the world of spirit and was intellectually predisposed to begin with truth and work toward the significance of facts. Thoreau, on the other hand, was an idealist who deeply delighted in the actual world, particularly the wilderness, and was predisposed to begin with facts and study how they flowered into truth. Emerson's paradigmatic "transparent eyeball" experience, described in the most memorable passage of *Nature*, occurred while crossing Boston Common; Thoreau's analogous "Contact" experience took place on the burnt over slope of a remote mountain in the wilds of Maine. These fundamentally mystical experiences occurred in locations and were described in prose that highlights the differences between the two men's interests and their approaches to knowledge.

Between the fall of 1849 and the fall of 1850, Thoreau dramatically reoriented his personal, professional, and artistic life in ways that indicate clearly his resolve to focus on developing his own distinctive interests and methodologies. Basically, he had been an Emersonian transcendentalist; in 1849, he resolved to become a Thoreauvian transcendentalist. He adopted a daily routine of morning and evening study and writing separated by an extended afternoon excursion into the countryside around his hometown; earned a livelihood by periodically surveying his neighbors' woodlots; and began intensive, formal studies of natural history, particularly botany, and of aboriginal cultures, especially those of North America. During the latter part of this transitional period, he began dating each entry in his journal and stopped cutting pages from the journal as part of his composition process: both changes ensured the scientific integrity of the data he collected during his afternoon excursions. Finally, in late 1850, he began assembling passages from his journal for a lecture on the topic that he had decided would occupy him for the remainder of his creative life—nature, particularly wilderness and that part of nature which he called "wildness" or "the wild." Because "Walking," the essay version of this lecture, did not appear in print until a month after his death, it is easy to think of the essay as his final word on wilderness when it is actually a midcareer summation and, as Thoreau himself referred to it, "a sort of introduction to all" he planned to write after about 1854, when *Walden* was published.[17]

He delivered "Walking" in the vestry of Concord's Unitarian Meetinghouse on April 23, 1851. After taking as his "rallying cry" two lines from Wordsworth, the English nature poet, about "stepping westward" and after apologizing for not speaking about the recently passed Fugitive Slave Law, he told his townspeople:

> I had prepared myself to speak a word now for *Nature*—for absolute freedom & wildness, as contrasted with a freedom and culture simply civil—to regard man as an inhabitant, or a part and parcel of nature—rather than a member of society. I wish to make an extreme statement, if so I may make an emphatic one, for there are enough champions of civilization—the minister and the school committee—and every one of you will take care of that.

He addressed his audience that evening a half mile from Old North Bridge, where seventy-six years before a group of Concordians had begun the revolution that led to the civil freedoms to which he referred in his opening remarks. "Walking" was his opening salvo in a new, far more radical revolution to extend freedom beyond a merely civil frame of reference to a truly universal and absolute one.[18]

Henry Thoreau fired his "shot heard round the world" that April evening in 1851 by famously asserting, "in Wildness is the preservation of the world." Atop

Katahdin, he had witnessed pure matter being transmuted into the raw material of life. In "Walking," he suggests that matter becomes the raw material of life by the infusion of wildness, which may be defined as the primal life force of the universe. Once infused into matter, wildness is recycled in nature's eternal return of life, death, and rebirth—a recycling which literally preserves the physical world. In the form of various natural resources, civilization draws wildness from the wilderness in order to fulfill civilization's many beneficial ends, the most fundamental of which is to insulate its inhabitants from the physical dangers of the wilderness. But in insulating its inhabitants from those dangers, civilization risks denying its inhabitants the moral and intellectual benefits available from the wilderness. One such benefit is a broadening of conceptions about humanity and our place in the universe. Conceptions developed within exclusively civilized frameworks generally and perhaps invariably reflect their insular, anthropocentric origin by falsely regarding humans as superior to or otherwise separate from nature. How might individuals maximize the benefits of both civilization *and* wilderness while minimizing the dangers and shortcomings of both? Thoreau provides a convincing answer to this important question.[19]

He describes wildness as a bracing tonic for humans. The restorative, therapeutic effect that wildness has upon us arises in significant part from our witnessing instances of life feeding upon death and being reborn. Deliberating upon these and other manifestations of wildness prompts us to realize our limitations, the most fundamental of which is our mortality. We, too, shall die one day and return our mite of wildness to nature's eternal return. Witnessing our own limits transgressed helps us to understand that we as human beings are not the measure of all things, that we are simply part and parcel of the infinite and eternal universe. The wilderness, which is of course where wildness is most in evidence, is thus an unparalleled proving ground, the best place for humans to test our mettle, because nowhere else can we get civilization's relatively trivial, myopically anthropocentric concerns so entirely out of our minds, leaving us free to develop the widest possible views of the universe, humanity, and our true place in the universe. In "Walking," Thoreau describes a place "where one primitive forest waves above, while another primitive forest rots below" as being "fitted to raise not only corn and potatoes, but poets and philosophers for the coming ages." Again, intellectually reflective contact with wildness as manifested in those living and rotting forests produces wisdom, a true conception of our condition (*who* we are) and the infinite extent of our relations (*where* we are). In wildness, therefore, is not only "the preservation of the world"; in wildness is also the salvation of human beings living in this world. Within a month after delivering "Walking," Thoreau wrote in his journal, "How important is a constant intercourse with nature and the contemplation of natural phenomena to the preservation of moral and intellectual health. . . . He approaches the study of mankind with great advantages who is accustomed to the study of nature."[20]

More than two years after delivering "Walking," Thoreau again addressed his townspeople and added a significant preservationist element to his evolving conception of the wilderness. In "Walking," he had pointed out that, "with regard to Nature," he lived "a sort of border life" between the wild and the civilized; in December 1853, he amplified this point by suggesting that poets and philosophers, seekers of beauty and wisdom, would do well to live in villages bordered by landscapes that are "the natural consequence of what art and refinement we as a people have," landscapes which he described more particularly as "woods and fields . . . [with] primitive swamps scattered here and there in their midst, but not prevailing over them." Such landscapes enable poets and philosophers to enjoy the advantages of both wilderness and civilization. Importantly, though, Thoreau insisted that poets and philosophers must from time to time leave their villages and bordering landscapes in order to travel "far in the recesses of the wilderness" where they can "drink at some new and more bracing fountain of the Muses." To satisfy this periodic but nonetheless critical necessity for more intense contact with the bracing tonic of wildness, large tracts of wilderness, "our national preserves," must be set aside, "not for idle sport or food, but for inspiration and our own true re-creation." This 1853 lecture, originally published in 1858 and reprinted later as the "Chesuncook" chapter of *The Maine Woods*, is one of our earliest, most eloquent statements on the need to preserve wilderness. For Thoreau, just as wildness preserves the material world, wilderness inspires highly cultured individuals to recreate themselves, to realize with renewed emphasis who and where they are. With that renewed sense of themselves, they are able to satisfy the transcendentalist imperative to "enjoy an original relation to the universe."[21]

Thoreau spent the remainder of his life working to put scientifically viable foundations under this distinctively Thoreauvian constellation of wilderness-inspired ideas. As early as December 1837, he had written, "How indispensable to a correct study of nature is a perception of her true meaning. The fact will one day flower out into a truth." By 1851, he had a strong sense of nature's true meaning, thanks in large part to Emerson. He now recognized that he would need to master a scientific study of nature so that he could follow and articulate the steps by which fact flowered into truth. This recognition accounts for his dramatic reorientation of 1849–1850. He tasked himself with mastering a science that aspired to learn not simply the laws of the physical universe, but the truth of Emerson's seemingly bifurcated universe of spirit and matter—"a science," to use Thoreau's own words, "which deals with the higher law."[22]

His decade-long odyssey began in earnest with a reassessment of the data he had gathered while sounding Walden Pond during the late winter of 1845–1846. He brought to that reassessment an exciting insight extrapolated from reading an appendix to the 1848 annual report of the Smithsonian Institution, wherein meteorologist Elias Loomis employed clearly articulated scientific principles to conclude, "When we have fully learned the laws of storms, we shall be able to *predict* them. This attainment is of the highest practical importance." For Thoreau, fully learning natural laws provided a jumping-off point toward an understanding of their importance, for he was intent on learning their moral and intellectual import for all humanity, not simply their "practical importance," which Loomis suggested was their "value to the farmer" and navigator. This ambitiously expanded mandate of deriving moral and intellectual truths from natural facts was the only difference between Thoreau and more traditional scientists. He added his Loomis-inspired extrapolation to the *Walden* manuscript, probably in 1852:

> If we knew all the laws of Nature, we should need only one fact, or the description of one actual phenomenon, to infer all the particular results at that point. . . . The particular laws are as our points of view, as, to the traveller, a mountain outline varies with every step, and it has an infinite number of profiles, though absolutely but one form. Even when cleft or bored through it is not comprehended in its entireness.[23]

This extrapolation springs from Loomis's scientific insight to assert that human knowledge is never and can never be sufficient, a principle that appears in "Walking," where Thoreau states, "The highest that we can attain to is not Knowledge, but Sympathy with Intelligence." If the universe were simply physical and finite, humans might hope eventually to know "all the laws of nature," but an entirely comprehensible universe would not satisfy what Thoreau regarded as humanity's need for infinite wildness and mystery, our need "to witness our own limits transgressed." Indeed, in *Walden*, Thoreau all but defines wilderness as that portion of the physical universe which at any given moment remains unexplored and unknown. Because the universe is infinite (and, Thoreau and Emerson assert, spiritual as well as material), it accommodates both humanity's need for mystery *and* our earnest desire "to explore and learn all things."[24]

Although Thoreau's science of the higher law aimed beyond knowledge to "sympathy with intelligence," it enabled him to generate an impressive amount of viable scientific knowledge because it employed precisely the same rigorous methods as standard science. He studied botany during the early 1850s at least in part because he aspired to become intimately acquainted with one of nature's most obvious cyclical phenomena, the seasons. On April 18, 1852, after weeks of tirelessly observing and recording the multifarious phenomena of spring, he wrote in his journal, "For the first time I perceive this spring that the year is a circle. I see distinctly the spring arc thus far. It is drawn with a firm line." His

Walden soundings had reflected "a rule of the two diameters," which he applied successfully to approximate White Pond's deepest point, just as Loomis expected to "predict" storms once meteorologists had fully learned their laws. Now Thoreau's seasonal data on the spring began to reflect another telltale pattern. But the next paragraph in his journal shows his impulse to make his facts flower *beyond* natural laws:

> Why should just these sights and sounds accompany our life? . . . I would fain explore the mysterious relation between myself and these things. I would at least know what these things unavoidably are, make a chart of our life, know how its shores trend—that butterflies reappear and when—know why just this circle of creatures completes the world. Can I not by expectation affect the revolutions of nature—make a day to bring forth something new?

Just as he had extracted ethical significance from the chart of his soundings of Walden and the trending of the pond's shores, he amassed seasonal data so that he would be able to participate creatively, "by expectation," in nature's regenerative cycles.[25]

He continued gathering seasonal data in Concord as long as his health permitted, periodically gleaning data from his journal, arranging them in chronological lists, and then moving the data into the cells of large charts with the years along the top and seasonal phenomena along the left margin. These phenological charts are one of several large projects carried on by this ambitious natural philosopher. His nature studies remained intense throughout the 1850s, in both the wilderness around Concord and his attic study. He read widely and keenly, eventually compiling in notebooks almost four thousand pages of extracts from books of natural history, indigenous cultures, and wilderness travel. And he spent at least four hours each afternoon in the fields and woods, recording his observations in a field notebook so that he could expand them later in his journal. Studying nature so closely and with such a unique perspective, he developed within an amazingly brief period an uncannily prescient sense of nature's vast cycles and vital interrelationships. Because he also developed a rigorously scientific understanding of those interrelationships, he well deserves the recognition conferred upon him as one of our first true ecologists.[26]

The most compelling indication of Thoreau's scientific sophistication toward the end of his life is his remarkable response to *On the Origin of Species*. He was one of the first Americans to read Darwin's great synthesis—and, in part because of his wide and careful reading, perhaps the only American who understood the subtleties of Darwin's argument *and* embraced that argument enthusiastically, without expressing a single reservation. Indeed, he found Darwin's evolutionary world view as encouraging as Malthus's strangely economic one, although for entirely different reasons. Darwin's "development theory," as it was then called, implied to Thoreau "a greater vital force in Nature, because it is more flexible and accommodating, and equivalent to a sort of constant new creation." Emerson had asserted that a creative force informs and directs matter, a force which seems analogous to the mysterious powers atop Katahdin that create the raw material of life by infusing wildness into pure matter. Thoreau apparently believed that Darwin's theory supported, or at least did not conflict with, Emerson's assertion and, by implication, Thoreau's own ideas about wildness.[27]

Thoreau also responded to Darwin by working to complement the "geographical distribution" component of the English naturalist's evolutionary argument with Concord-specific data on seed dispersion and forest succession. Darwin had confessed the scientific community's general ignorance of seed-dispersal mechanisms, and Thoreau hoped that if he clearly explained the principles of tree succession to his neighbors, they might adopt silvicultural practices that worked with rather than against nature. He delivered "The Succession of Forest Trees" in September 1860 but did not live long enough to complete *The Dispersion of Seeds*.[28]

The longer and more closely Thoreau studied nature during the 1850s, the less bifurcated his view of the universe became. A sentence in his journal of April 1859 suggests how far he had traveled since his Cohasset visit of 1849: "There is no other land; there is no other life but this, or the like of this." That final qualifier marks him as an unregenerate idealist. The trajectory of his conceptions—from his earlier Emersonian belief that people's spirits depart to "a newer world" after death to this later assertion that humans can look forward to no other world or life than "this, or the like of this"—suggests that he would have continued to study nature's dynamics until he could confidently assert that heaven is *literally*, as he phrased it in *Walden*, "under our feet as well as over our heads." Tragically, though, America's first apostle of wilderness died of tuberculosis on May 6, 1862, two months before his forty-fifth birthday. At Thoreau's funeral, Emerson alluded to *The Dispersion of Seeds*, the phenological charts, and his late friend's other uncompleted wilderness-based projects as a "broken task which none else can finish," rightly surmising that the "scale on which [Thoreau's] studies proceeded was so large as to require longevity."[29]

Understanding that Thoreau engaged in his detailed scientific study of nature's phenomena within the larger framework of transcendentalist assumptions about the universe, we can appreciate the incredible scope of his ambition. He developed during the course of a single decade a highly sophisticated, truly ecological understanding of wilderness and the rest of the physical world. But his conception of the universe as a whole, while embracing the scientific method as a means of knowing the physical universe, remained fundamentally romantic by being responsive to humanity's requirement "that all things be mysterious and unexplorable, that land and sea be infinitely wild, unsurveyed and unfathomed by us because unfathomable."[30]

Notes

1. Nathaniel Morton, *New England's Memorial*, 5th ed., ed. John Davis (Boston: Crocker & Brewster, 1826), 35; Henry D. Thoreau, *Journal*, vol. 3, *1848–1851*, ed. John C. Broderick et al. (Princeton, N.J.: Princeton University Press, 1990), 27.

2. "About the APS," http://www.amphilsoc.org/about (accessed 7 December 2004); Cotton Mather, *The Christian Philosopher* (London: Eman. Matthews, [1720]), 3.

3. Pehr Kalm, *En Resa till Norra America* (Stockholm, Sweden: Lars Salvi, 1753–1761).

4. Thomas P. Slaughter, *The Natures of John and William Bartram* (New York: Knopf, 1997).

5. [Edmund Burke], *A Philosophical Enquiry . . . on the Sublime and Beautiful* (London: R. and J. Dodsley, 1757), 132–33.

6. William Bartram, *Travels through North and South Carolina* (Philadelphia: James & Johnson, 1791), xxiv–xxv, 251, xxxiv, xxiv, 343.

7. Henry D. Thoreau, *Letters to a Spiritual Seeker*, ed. Bradley P. Dean (New York: Norton, 2004), 256–57, n. 10.

8. Walter Harding, *The Days of Henry Thoreau* (New York: Knopf, 1962), 10; *The Writings of Henry D. Thoreau*, ed. Bradford Torrey and Francis H. Allen (Boston: Houghton Mifflin, 1906), 8:306–7, 7:296 (hereafter cited as *W*). This edition is gradually being superseded by *The Writings of Henry D. Thoreau* (Princeton, N.J.: Princeton University Press, 1971–), fourteen volumes to date (December 2004). Hawthorne, quoted in Harding, *Days of Henry Thoreau*, 146–47.

9. Ralph Waldo Emerson, *Nature* (Boston: James Munroe, 1836), 5, 8.

10. Emerson, *Nature*, 80.

11. *W* 3:70–71.

12. *W* 3:77–79.

13. Ian Marshall summarizes the extensive literature on the "Contact" passage controversy in *Story Line: Exploring the Literature of the Appalachian Trail* (Charlottesville: University Press of Virginia, 1998), 226–48; *W* 6:319–21.

14. *W* 2:190.

15. *W* 2:349–50, 12, 350.

16. *W* 4:6–7, 11–13.

17. Bradley P. Dean, "A Sort of Introduction," *Thoreau Research Newsletter* 1, no. 1 (January 1990): 1–2.

18. Bradley P. Dean and Ronald Wesley Hoag, "Thoreau's Lectures before *Walden*: An Annotated Calendar," in *Studies in the American Renaissance 1995*, ed. Joel Myerson (Charlottesville: University Press of Virginia, 1995), 198–200.

19. *W* 5:224.

20. *W* 5:229, 8:193.

21. *W* 5:242, 3:172–73.

22. *W* 7:18, 11:4.

23. Elias Loomis, "Appendix No. 2. Report on the Meteorology of the United States," in *Report of the Board of Regents of the Smithsonian Institution*, 30th Cong., 1st sess., 1848, S. Misc. Rept. 23, 203; *W* 2:320.

24. *W* 5:240, 2:350.

25. *W* 9:438, 2:319–20.

26. Robert Kuhn McGregor, *A Wider View of the Universe: Henry Thoreau's Study of Nature* (Urbana: University of Illinois Press, 1997), 121–98.

27. On Thoreau and Darwin, see Bradley P. Dean, "A Textual Study of Thoreau's *Dispersion of Seeds* Manuscripts" (Ph.D. diss., University of Connecticut, 1993), 8–14; Henry D. Thoreau, *The Dispersion of Seeds*, in *Faith in a Seed*, ed. Bradley P. Dean (Covelo, Calif.: Shearwater Books/Island Press, 1993), 102.

28. Bradley P. Dean and Ronald Wesley Hoag, "Thoreau's Lectures after *Walden*: An Annotated Calendar," in *Studies in the American Renaissance 1996*, ed. Joel Myerson (Charlottesville: University Press of Virginia, 1996), 339–47. The manuscript pages of *The Dispersion of Seeds* were reconstructed and published in 1993 (see n. 27 above).

29. *W* 17:159, 2:313; Joel Myerson, "Emerson's 'Thoreau': A New Edition from Manuscript," in *Studies in the American Renaissance 1979*, ed. Joel Myerson (Boston: Twayne, 1979), 55, 54.

30. *W* 2:350.

Six

The Fate of Wilderness in American Landscape Art

The Dilemmas of "Nature's Nation"

Angela Miller

In the 2004 film *The Day after Tomorrow*, the Northern Hemisphere is engulfed by a new "ice age," a catastrophic climate change resulting from global warming that transforms the conditions of life on the planet. The administration in Washington, struggling to understand the scale of the crisis, meets in a rotunda somewhere in the White House, which is hung with four paintings by the leading artists of the nineteenth-century American landscape tradition, including Thomas Cole. No such space exists in the White House; however, the selection of landscape art as a backdrop to an unprecedented environmental crisis is entirely appropriate to evoke the central dilemma the film confronts: the interdependence of our advanced postindustrial society with a natural world whose laws it has consistently violated. The film's creators evidently banked on the symbolic resonance that grand images of the American wilderness continue to carry for American audiences as a symbol of a time when the nation was carpeted with forests instead of highways and factories, preserving an equilibrium crucial to climatic and social stability.

The contemporary symbolism contained in these well-known works of landscape art is a muted echo of the even fuller public response they once provoked. Between the 1820s and the 1870s, the American landscape drew the fascinated attention of the nation's most accomplished painters and their large public audiences. Landscape painting as an expression of national identity first emerged in New York City. Promoted by wealthy New York patrons and supported by the leading cultural institutions of the nation's "empire" city (including the Century Club, the Union League, and the National Academy of Design), landscape painting spoke not only to the nation's cultural progress in the arts but also to its deepest ambitions as a republic.

By the peak of the genre's popularity in the 1850s and 1860s, leading landscape painters were celebrated figures, building lavish villas along the Hudson River as it wound its way down to New York. They would collectively come to be known as the "Hudson River School" even though their subject matter would eventually encompass a far larger area. What is most significant in the present context—the history of American ideas about wilderness—is that these wilderness ideas were conveyed to the public by *representations* of nature, that is, paintings, more powerfully than by the real thing itself. Eastern Americans were more likely to see their wilderness in a gallery than on an expedition. And viewing these images was increasingly structured around shared public meanings, grounded in well-understood symbols. A nation of Bible readers, for instance, recognized rainbows in art as a sign of providential grace bestowed on those carrying out God's work. Giving weight to this public symbolism were crowds of people poised before spectacularly detailed images, opera glasses in hand, and assisted by written explanations more than twenty pages long in the case of Frederic Church's "Heart of the Andes."

Images of the American landscape extended far beyond paintings on canvas. Popularized in such large-scale publications as the two-volume *Picturesque America* and celebrated in verse and travel literature, landscape representations were used to encourage tourism. Images of the American landscape were also among the first popular native expressions of cultural nationalism in the early decades of the nineteenth century; the United States turned to Romantic images of nature as a source of patriotism. Anglo-Protestant beliefs merged with wilderness ideas and shaped the United States' emergent sense of exceptionalism: the idea that America was different than Europe because of its nature, a place apart, an unpeopled wilderness where history, born in nature rather than in corrupt institutions, could begin again. As shown by Mark Stoll in chapter 3, "sublime" wilderness was a place where moral and spiritual virtue would be renewed, where God spoke to his new "chosen."

Such beliefs, however, left considerable space for cultural debates over the particular relationship of wilderness to national culture. The paradox of landscape as a subject of art and a source of national pride was that it arose at a time when the United States was busily occupied in converting those same landscapes into commodities through industry and market capitalism. In this chapter, I will look more closely at three distinct aesthetic responses to the dilemma of "nature's nation": how, that is, to reconcile wilderness as the nation's birthright and unique heritage with economic and social development as the nation's imperative. These three case studies—the wilderness ideal of critical Romanticism, the middle landscape ideal of harmony between nature and culture, and the turn to preservation through federal protection (setting undeveloped nature apart from development)—follow a rough historical progression extending from the early to the later nineteenth century. Any understanding of the meaning of wilderness in the nineteenth century must take account of this changing intellectual, aesthetic, and social history.

Critical Romanticism and the American Landscape

A cultural, artistic, and philosophical movement of international reach, Romanticism developed in part as a reaction to the environmental and social disruptions brought by the industrial revolution and market capitalism. Anglo-American Romanticism looked to nature as a source of personal renewal and as a pointed alternative to the disenchantments of modernity. Its vision of a world in which natural and human rhythms were harmonically attuned looked to the landscape for expression of interior moods and feelings.[1] These English and American Romantics challenged the extractive vision of nature as the source of raw materials with an aesthetic, philosophical, and spiritual commitment to wilderness—the wilderness ideal. Writers, poets, and painters from William Wordsworth to Emerson and Thoreau (explored in the previous chapter) attacked the utilitarian mentality behind capitalism. These obsessions, they felt, blinded society to the broader dimensions of a universe animated by natural energies, manifestations of a spiritual order far greater than the human. The Romantic wilderness ideal of an untouched, nonhuman, timeless source of moral authority was the invention of a particular historical moment, expressing longing for an alternative to an entirely human-centered world.[2] Pursuing this vision, Romantics questioned the consequences of imperious attitudes toward the conquest of nature, and they would increasingly come to see such attitudes as a critical challenge to the future health of society.

One artist in particular—significant enough to be mentioned in the three preceding chapters—grasped the implications of the philosophical shift initiated by the culture of Romanticism. When Thomas Cole emigrated from England with his family in 1818, he brought to the United States a deeply religious response to his adopted country's embattled wilderness. Cole saw the nation's expansion across the continent as a tumultuous, destructive process that posed difficult, sometimes irresolvable dilemmas and choices for the new republic. It may seem strange that an Englishman was the first artist to express a Romantic vision of the American wilderness as a powerful spiritual and national resource, but Cole's family had come from Lancashire, a region of England that had early on felt the full force of the industrial revolution's blight on the rural districts of England. Cole thus carried an intensified awareness of natural fragility and of

FIGURE 6-1. Thomas Cole, *Landscape* (1828). Museum of Art, Rhode Island School of Design, Walter H. Kimball Fund (30.063).

the destructive powers of industry, enhanced by his familiarity with Romantic literature. Largely self-taught in painting, Cole more than any other artist of the time instinctively grasped the dramatic potential of the Catskill Mountains just north of New York City. In no sense a wilderness, the Catskills harbored a corrosive tanning industry that produced wide deforestation. It was, in addition, a popular tourist destination.

Cole's Catskill paintings of the 1820s, however, as well as those of the White Mountains in New Hampshire, spoke eloquently to his metropolitan audience of a land fresh from the hand of Providence, energized by the cycles of the seasons and the rhythms of natural processes: life, decay, and rebirth (see fig. 6-1: *Landscape with Tree Trunks*). While confident of their position at the helm of a growing commercial and trade empire, particularly after the opening of the Erie

Canal in 1825, these audiences were often only one generation removed from a rural existence. They received a shock of recognition on seeing Cole's first landscapes, which were exhibited in a New York shop window in 1825. Earlier landscape art had been topographical in nature, tamely delineating estates and property, rather than "compositions" highlighted by expressive natural features and dramatic weather. While still in his twenties, Cole won major patronage and critical acclaim.

But from the start, the cult of wilderness diverged in troubling ways from everyday behavior. The audiences of Cole's Catskill paintings could see his Romantic wilderness as the treasured symbol of America's exceptionalism. Yet these same audiences were deeply implicated in the market revolution that was rapidly transforming the metropolitan hinterlands, through their activities as entrepreneurs and businesspeople. Landscape art offered a therapeutic retreat from the forces of market development in which they themselves were involved, forces that were endangering the very wilderness to which they turned for refuge. Cole saw this dilemma more clearly than most: the wilderness that guaranteed America's privileged conversation with God and which was central to America's emerging identity as a republic was under attack by Americans themselves.

Cole's ambitious five-part cycle of paintings *The Course of Empire* most fully explored the link between the fate of nature and the future of the republic. It did so, however, in an allegorical form that left unnamed the subject of his moral tale—at first glance, a republic in the ancient world. The first three canvases of *The Course of Empire* told a story familiar to Americans in the 1830s—the rise of a great empire from origins in primitive wilderness. In the first canvas, *The Savage State* (fig. 6-2), hunters wearing animal skins roam through a feral landscape of forest and mountains as mists rise from the sea, suggesting the infancy of culture. The second, *The Pastoral or Acadian State*, shows a domesticated nature harmoniously poised between wilderness and civilization, a momentary balance upset in the third and central canvas of the series, *Consummation* (fig. 6-3), a glittering image of a maritime empire. *Consummation* carried a familiar lesson for American audiences, long taught to distrust too much wealth and luxury as leading to moral and political corruption (exemplified by the central image of an emperor held aloft like a god). The arts, which in the previous canvas seem to grow gracefully out of nature, now appear monstrously excessive. The place-defining mountain peak of the first two canvases has virtually disappeared beneath the elaborate architecture of empire. *Consummation* sets the stage for the final two canvases, which play out the implications of imperial arrogance. In *Destruction*, an invading army overwhelms the empire, recalling the sack of Rome. The violence of the invaders, however, merely acts out the underlying ruthlessness of the empire itself, in its exploitation of nature. Cole concluded the

FIGURE 6-2. Thomas Cole, *The Course of Empire: Savage State* (1836) (first in series). Collection of the New-York Historical Society (acc. no. 1858.1).

series with *Desolation*, a haunting moonlit image of the empire in ruins, which returns the nation to its beginnings in nature.

Cole's *The Course of Empire* brought to light in striking fashion an anxiety shared by many of his contemporaries: in the words of historian Perry Miller, the "secret, hidden horror that its gigantic exertion [of American empire building] would end only in some nightmare of debauchery called civilization."[3] Such a vision of historical defeat, Miller suggested, might seem incongruous for a nation confident of its Christian civilizing mission and its future greatness. Audiences of Cole's series did in fact resist its possible meanings for the American republic of the 1830s, but Cole's letters and papers suggest that he was thinking deeply about his adopted country. *The Course of Empire* was an object lesson that graphically revealed the catastrophic results of falling away from nature. Republics, as any student of antiquity would know, stand or fall on the virtue of their citizens, and what guaranteed that virtue in the United States was proximity to wilderness (a primary tenet of cultural nationalism in these decades). Starting with the premise of America's historical exceptionalism—its ability to escape the laws of cyclical rise and decline that defined the empires of the Old World—such logic pointed to an unavoidable conclusion: as the nation's forests

FIGURE 6-3. Thomas Cole, *The Course of Empire: Consummation of Empire* (1836) (third in series). Collection of the New-York Historical Society (acc. no. 1858.3).

and unsettled regions receded before the onslaught of civilization, so did the source of its cultural virtue, leaving the fledgling republic vulnerable to the very debauchery so vividly imagined in the central canvas of Cole's series.

Cole's *View from Mount Holyoke, Northampton, Massachusetts* (known as *The Oxbow*) (fig. 6-4) was completed in 1836, the same year as *The Course of Empire*. Indeed, the works hold a revealing relationship to one another. *The Oxbow* offers an either-or scenario dramatizing two very different futures for the nation. A large canvas for the time (51½" × 76"), *The Oxbow* gave audiences a panoramic view of the Connecticut River valley in western Massachusetts where the river turns back upon itself, creating an unusual natural feature that drew tourists.

From the rugged vantage point of the nearby mountain, the valley below offers a vision of agrarian peace and plenty, all the more through contrast with the wilderness on the left side of the canvas. Such a reading would be consistent with the optimistic vision of those promoting the colonization of nature in the 1830s. This call for a domesticated nature was made according to the ideal of the middle landscape, which resolves the extremes of wilderness and civilization. The middle landscape encouraged the idea that the nation could enjoy progress

FIGURE 6-4. Thomas Cole, *View from Mount Holyoke, Northampton, Massachusetts, after a Thunderstorm (The Oxbow)* (1836). Metropolitan Museum of Art, Gift of Mrs. Russell Sage, 1908 (08.228).

and economic development without severing its ties to nature. Cole's *The Course of Empire,* however, confronted the impossibility of realizing the middle landscape in real historical time. In different ways—a five-part serial composition and a panoramic view that implies extension beyond its frame—Cole insistently located the middle landscape and the wilderness within a broader trajectory of change and development. Weaving Romantic artist, nature, and nation into a dynamic image of environmental change, *The Oxbow* was fraught with troubling implications for the future of the republic.

In a letter to his patron Luman Reed, Cole wrote that in his painting of the Oxbow, he wished "to tell a tale."[4] Though he did not reveal the tale to be told, the painting strikes viewers with its narrative power. Indeed, its panoramic breadth links it to the popular genre of the 360-degree stationary panorama, the diorama, and the moving panorama, which was painted on long strips of canvas and then unrolled across a stage, providing an explicitly narrative dimension as the view passed before the audience.[5] The impulse to read *The Oxbow* panoramically is encouraged by the storm front that engulfs the left, or wilderness, side of the landscape. Cole's *Oxbow* collapsed two of the first three stages of empire: wilderness and pastoral nature. Yet reading the canvas laterally, the implication is that the process of development will carry us from agrarian pastoralism toward increasing settlement and urbanization. History—his painting implies—cannot be stopped.

Cole left little doubt that he intended his tale to transpose the allegory of ancient empire, dramatically told in his just-completed series, onto the young republic of the 1830s. On the distant hillside that breaks the horizon, its mount just brushed by the advancing storm front (or is it retreating?) is a series of markings that have been read by scholars as Hebrew letters for the "Almighty."[6] Cole was convinced that his adopted nation had a providentially appointed mission of redemption to fulfill—America, as he wrote elsewhere, was a new Eden. Yet the nation's privileged position in history was threatened by the ignorance and greed of its own citizens and by an unquestioning devotion to development. This message, couched in natural terms but coming from beyond nature, points toward an open-ended future in which confidence about the direction of the nation gives way to a prospect as unstable as the weather itself. Nature here seems to act out the ambivalence felt by many of Cole's contemporaries.

The conflict between Cole's Romantic devotion to wilderness and the imperial conquest of nature was intensified by the agrarian character of the republic before the Civil War. While manufacturing and industry certainly played a role in transforming the countryside, farming wrought far greater damage to the forests of North America, as Steven Stoll has noted in chapter 4. Practices such as slash-and-burn agriculture and tree girdling (removing enough of the bark to kill the tree), which were commented on by numerous European visitors to the New World, assaulted aesthetic values rooted in eighteenth-century categories of the beautiful and the picturesque, that is, harmony between part and whole, smooth spatial transitions, and a modulation between open fields and forests. Basil Hall, an Englishman who had come to the United States in the late 1820s to secure sketches of the scenery for publication at home, wrote of newly cleared lands as having "a bleak, hopeless aspect . . . cold and raw," and lacking the settled, aesthetically pleasing appearance of European nature. Such scenes, he concluded, had no parallel in the Old World.[7] Slash-and-burn and girdling practices served a rapidly expanding nation of farmers who pitted themselves against a "wilderness" that required taming in order to yield economic profit or even subsistence. With few exceptions, American artists avoided such scenes of aesthetic and environmental devastation, preferring the comforting fiction of the "middle landscape" to the realities of an unsettled nature in transition. A notable exception is Asher B. Durand's *First Harvest in the Wilderness* (1855, fig. 6-5), an

FIGURE 6-5. Asher B. Durand, *The First Harvest in the Wilderness* (1855). Brooklyn Museum. Transferred from the Brooklyn Institute of Arts and Sciences to the Brooklyn Museum.

unusually frank representation of the scars of deforestation that accompanied the establishment of a yeoman's empire. Yet Durand used painterly means—a softening atmospheric haze, illuminated by filtered sunlight, and a refined gradation of muted tones—to blunt the harsh edges of a landscape undergoing dramatic change.[8]

Cole was acutely aware of the devastating impact of agriculture. Images of natural desolation, of "prostrate trees—black stumps—burnt and deformed," recur throughout his journals. English Romantic that he was, Cole experienced such injuries to nature on a deeply personal level, associating them symbolically with the "wasted places" of the American spirit. The barrenness of nature held for him the threat of artistic impotence; he saw colonization as a process by which nature's energies—tied to his own creative power—were drained away. The artist, bereft of the spiritual and aesthetic resources of wilderness, would have nowhere to turn for spiritual and creative renewal. Environmental destruction, motivated by the quest for economic gain, threatened cultural sterility.[9]

Cole's concerns proved prophetic; by midcentury, many others shared his alarm over the impact of "Yankee enterprise" and the "axe of civilization."

Indeed the complaint that America's wilderness was passing into the mists of history, like its human counterpart, the Native American, had lost force through overstatement. But few matched the moral conviction of Cole's attack on American utilitarianism. And his broader vision of the dire impact of American settlement on nature would not be equaled until 1864, with the publication of George Perkins Marsh's *Man and Nature* (see chapter 4).

Cole's visualization of the ravages of American farming anticipated Marsh's unblinking analysis of environmental devastation. Marsh's observations about the results of deforestation in the United States on drainage patterns and soil erosion formed only a small part of his global picture. From Marsh's perch in Italy, where he wrote *Man and Nature* while serving as minister to the newly unified nation, he was able to see his own United States within a broader history of environmental forces acting impartially across a range of geographical and natural conditions. Dispensing with notions of American exceptionalism, Marsh insisted that the only thing that distinguished the United States from Europe was its newness. He singled out agriculture—in particular, the cultivation of tobacco and cotton for export, along with domestic cattle—as most damaging to the American forest.[10]

Marsh's goal was what would later be termed *sustainability*, expressed with a prophetic grasp of the difficult choices his compatriots would have to make to achieve it. He summoned his contemporaries to put the welfare of future generations before their own "moral and material interests."[11] He redefined "civilization" as the realization of long-term environmental stability—the very antithesis of the pioneer ethos which "measured progress by the elimination of forests."[12]

Cole had no such developed program for reversing the environmental destruction of American "progress." There was, however, one moment in *The Oxbow* that implied that Americans could be more than passive witnesses to the historical processes transforming their landscape, that they could instead be decisive actors within history. Cole painted himself into the landscape. The artist, wearing a hat, sits before his easel; nearby is a pack and folded umbrella, evidence of the recent storm. Cole turns and looks at us, making us complicit with his act of representation and, by extension, moral witnesses to the changes in the land. To be a moral witness rather than merely a passive spectator is, however, to acknowledge one's role in history, with its burden of responsibility and choice. Three decades before Marsh, Cole turned away from the rhetorical bluster of the new nation-state, which was heedless of its fragile and finite nature, and brought dramatically to life the precariousness of American history and identity. Cole presented the middle landscape as necessarily transitory, its static harmony interrupted by the momentum of change sweeping across the republic like a storm front. Cole understood, as few of his contemporaries did, that the historical process was neither necessarily benign nor inevitable, but a product of social and moral decisions.

In the concept of wilderness, Cole enshrined all that differentiated the sensitive artist-poet and his aristocratic patrons from the unsavory new democratic energies driving the economic exploitation of nature in the new republic. What was true for Cole was also true for the Harvard historian Francis Parkman and other social and political conservatives. Parkman was from an old New England family that could afford to hold itself aloof from the self-making energies of those struggling for a stake in the material and social progress of the new nation (chapter 4's backwoods settlers). Traveling to the West in 1846 with his French guides, enjoying the manly solitude of the frontier while "reveling" in the poetry of the English Romantic Lord Byron, Parkman expressed contempt for the awkward, hungry-eyed men and women he encountered on the trail, who were driven by "the restless energy of [the] Anglo–American." His elitist disdain for those unable to rise above material or economic motives played out in his preference for "unaided nature" and for the natural grace of his guide, Henry Chatillon.[13] Parkman's *Oregon Trail*, his account of his journey to the West, plainly reveals the class-based nature of his preference for unsettled lands, free of the motley mix of humanity on the frontier and only lightly touched by the presence of Indians.[14]

By the 1850s, however, the ravaging energies of the pioneer—those locusts of the prairie who brought havoc in their wake—would be reenvisioned as part of an emerging mythology of the frontier. Western wilderness tested the self-reliance of a new democratic culture of self-made men, epitomized by the figure of the pioneer, building a future out of raw, unsettled nature. In works by Durand, Jasper Cropsey, George Caleb Bingham, and others, wilderness is the stage on which Americans enacted their historic destiny, valued not for itself but as a measure of the resourceful independence and fortitude of the American pioneer, who typified the new nation. The shift in the image of the pioneer, evident even in Cole's late work of the 1840s, was prophetic of a broader midcentury move away from the Romantic veneration of unsettled wilderness.[15] Yet these heroic backwoods settlers and land-hungry pioneers pursued destructive agricultural practices with dire environmental consequences.

From Criticism to Accommodation

Cole remained deeply pessimistic about the prospects for the young republic. By the 1840s, he had largely retreated from overt criticism and into an imaginary rural arcadia. Ironically, his critical message of the 1830s would be disarmed by artists such as Asher B. Durand, rising to prominence at midcentury, who increasingly modeled their work on the example of Cole's middle landscape,

FIGURE 6-6. Asher B. Durand, *Progress* (1853). From the Westervelt-Warner Collection of Gulf States Paper Corporation and on view in the Westervelt-Warner Museum of Art, Tuscaloosa, Alabama.

balanced between civilization and wilderness. Inspired by the pastoral second canvas of *The Course of Empire*, the midcentury middle landscape carried one important difference: it was lifted out of its context as one scene in a larger panorama of historical change, and instead transformed into an aesthetic formula whose major function was to disguise the most disturbing consequences of development. Affirmative rather than questioning, this version of the middle landscape circulated widely in the polite literature and art of the urbanized middle class.[16]

Among the most ambitious examples of the middle landscape at midcentury is Durand's *Progress (The Advance of Civilization)* (1853, fig. 6-6). In its panoramic breadth and symbolic ambitions, *Progress* recapitulated some of the historical themes that Cole had explored in his work of the 1830s. But Cole's cautionary attitude toward development is nowhere evident. Like *The Course of Empire*, *Progress* juxtaposes wilderness with the settled landscape, and like Cole's *Oxbow*, it too tells a tale. But Durand's tale now has replaced ambivalence and national reckoning with a fable of disarming simplicity. *Progress* moves smoothly through the stages in the transportation revolution, from wagon to canal to steamboat to train, the trestles of the railroad blending seamlessly into the contours of the landscape and its smoke dissolving without a trace into the atmospheric haze that blunts the raw edges of the developed landscape. Durand pushed the urbanized future—with its troubling implications for a culture long suspicious of industry and its effect on republican virtue—into the light-infused distance, where industrial fumes blend into morning mists. Unlike Cole, who used the pastoral aesthetic to emphasize by contrast the very *instability* of history, Durand's pastoral landscape seamlessly unites conflicting versions of the republic: nature's nation now is crisscrossed with train tracks and dotted with factories spewing smoke into cerulean skies, but all in a happy balance. Artistic skill here is directed at a form of cultural apologetics entirely different in tone from Cole's urgent summons to awareness.

Yet the story of *Progress* doesn't end there, for the painting contains hidden complexities. Witnessing the changes in the landscape are two Indians nestled in a lushly wooded foreground. Regret for the past (the unsettled wild landscape but also the intimate relationship to nature signaled by the Indians) mingles with embrace of the future in the theme of the "vanishing American Indian." Even as Durand celebrated the republic's dominion over nature, *Progress* is qualified by longing for what was perceived as the timeless and unchanging existence of those who lived in nature rather than acting upon it.

Unlike Cole's painting, which progresses laterally, Durand structured his landscape to be read from foreground into distance, the viewer moving deeper into the landscape. The Connecticut River of Cole's *Oxbow* turns back on itself to form a giant question mark, as if to punctuate the historical dilemma facing Americans and to call for national deliberation and choice. Durand's painting, full of visual blandishments, its surface carrying a buttery polish, offers certainty and resolution in the face of uncertainty and moral engagement.

What had changed? Public awareness of the destructiveness of settlement had, if anything, intensified since Cole's early embittered outbursts against those who cut down his beloved forests. Why then was Durand so quick to substitute a vision of smooth progress, tinged by only momentary regret for the threat such progress posed to American Indians? By the 1850s, the taste for wilderness had largely given way to a preference for the pastoral landscape. Sublime wilderness appealed less to metropolitan audiences than did the shared pleasures of a nature increasingly enjoyed for its parklike qualities, mirroring the middle-class discomfort with the intense, spiritually demanding nature of the Romantics. Eastern nature was now more clearly demarcated into areas of wilderness, settlement, and urban metropolis; these three geographies were, furthermore, seamlessly linked within an emergent network of markets extending outward from metropolis to hinterland. The result was that the contested status of nature between wilderness and settlement had been, for the moment at least, safely adjudicated by the establishment of a metropolitan market culture that wove these distinct arenas into a smooth fabric serving the increasingly confident nation-state at midcentury.[17] The problems of the wilderness were displaced to the frontier West. And in the landscape painting of the newly opened West, as we will see, artists employed many of the same strategies and aesthetic solutions after the Civil War to negotiate the stress of conflicting ideals: wilderness and habitat preservation versus a powerful extractive ethos that viewed nature as raw material for a postwar society increasingly defined by industry, transportation, and national wealth.

Yet in this ongoing effort to find a place for nature within an increasingly populous and powerful nation, few acknowledged Native American claims on the land. Like the forests of North America, their presence lay directly in the way of the republic's imperial ambitions. The idea of a wilderness untouched by human habitation hardly described the actual historical circumstances of a land not just inhabited but transformed and adapted to human needs for millennia. Since the 1980s, environmental historians have begun to grasp the extent to which the concept of wilderness erased the long-standing history of native cultures in the New World and ignored the ways in which these cultures had remade nature and created new ecological regimes, as demonstrated by Perreault in chapter 2. As visions of nature, both the Romantic wilderness and the pastoral aesthetic of the middle landscape ignored the painful social problem of indigenous people whose lands and ways of life were directly challenged by social and economic expansion.[18]

Wilderness Preservation and Native Dispossession

In 1872, seven years after the conclusion of the Civil War, Thomas Moran wove together the complex geological details of a sublime natural site into a grand synthesis that revealed to the curious gaze of easterners the mysteries that lay "at the heart of the continent." The immediate occasion for his painting *Grand Canyon of the Yellowstone* (fig. 6-7) was the survey expedition of Ferdinand Vandiveer Hayden in 1871, one of a series of such federal explorations of the continental interior launched by the newly formed U.S. Geological Survey to prepare the West for railroads, settlers, and extractive industry. The national park system was created in the midst of this massive federal investment in a region that would play a central role in the further industrial development of the nation-state.

In the decades following the Civil War, the nation's relationship to its wilderness had changed directions once again, producing the first official

FIGURE 6-7. Thomas Moran, *The Grand Canyon of the Yellowstone* (1872). Smithsonian American Art Museum, lent by the Department of the Interior Museum (L.1968.84.1).

acknowledgment that wilderness was an embattled condition that required federal protection. Unlike the earlier Romantic concept of Cole—in which wilderness's worth was intrinsic and linked to the individual spiritual bond with nature—the postwar American ideal of wilderness now fully embraced the concept of the powerful nation-state. The wilderness preservation movement and the national park system offered a resolution of sorts to the prewar dilemma of nature's nation by setting aside areas of nature as national shrines protected from development. But the creation of wilderness preserves after the war once again failed to accommodate the Native Americans who had long occupied these lands. Indeed, the institution of a national park system revealed the underlying problem at the heart of the wilderness ideal itself: its refusal or inability to accommodate the human presence, even when this presence had been an integral part of the very wilderness being protected from it. Ben Johnson will take this theme further in the next chapter, while I focus upon the role of artistic representation in this process.[19]

In 1872, the same year that Moran completed his enormous painting, the U.S. Congress designated Yellowstone as the nation's first national park. Moran's earlier watercolors of the region, circulated to members of Congress in a form of visual lobbying, played an instrumental role in the move to preserve Yellowstone as a "wilderness." Like Thomas Cole, who had initiated the American Romantic landscape tradition a half century earlier, Moran was an Englishman from Lancashire. And like Cole, he grasped what would spark his audiences to a shock of recognition, this time before the grand spectacle of a minutely detailed and sublimely scaled nature. The purchase of *Grand Canyon of the Yellowstone* by Congress marked the beginning of an artistic career dedicated to painting radiant scenes of the sublime western landscape. This sublimity, however, omitted all signs of railroads, mining, and settlement. What made it all the more compellingly real was that the final work was built up from field sketches and watercolors, often highly detailed line drawings that "mapped" the landscape. These revealed the influence of English critic John Ruskin, who insisted on the authority of natural truth and the importance of study from nature. But this new deference to nature still went hand in hand with persistent cultural narratives of colonization, as if there were no conflict.

Moran's magisterial image shows Yellowstone Falls exhaling a great column of spume into the sky above, while the aquamarine of the river below traces its course through a graceful V-shaped valley. Standing on a rocky promontory surveying the scene in the middle foreground are two tiny figures: a hatted man gesturing toward the scene before him, representing the figure of Ferdinand Vandiveer Hayden, leader of the scientific survey team into the Yellowstone, and a second man dressed in the ceremonial fashion of a Native American chieftain. Around his neck he appears to be wearing a medal of the sort given by the federal government to native groups in acknowledgment of treaty agreements over land transfers. The two men are positioned around a common axis formed by a spear-like pole; the native figure faces the viewer, his back turned to the landscape, while the surveyor faces toward the great valley. Together, they represent past and future, at a moment of symbolic transfer when the mysterious grandeur and untold economic, scientific, and social wealth of the West changed hands from native inhabitants to the federal government.[20] Countering the legendary image of Yellowstone as an infernal region of dangerous energies, a place resistant to human presence, Moran organized the scene to conform to aesthetic ideals of symmetry. His balanced composition alternates shadowed foreground with light-filled middleground, and a darker distance with the point of greatest visual emphasis and highest value—the falls themselves. This impressively orchestrated scene combines breadth of vision with depth of geological detail; vision here serves the broader objective of social and scientific understanding and, ultimately, control of the West.

Grand Canyon of the Yellowstone grandly synthesized a vision of the western wilderness as a national shrine protected from the very forces of development to which Moran himself was indirectly allied as the artist accompanying the Hayden expedition.[21] Vastly enlarging the panorama of nature in the paintings of Cole and Durand (*Yellowstone* measures 7' × 12'), Moran's painting represented a third and final heroic phase of American landscape art as an expression of national identity. By the 1870s, however, the symbolic uses of wilderness had changed once again. The spiritual and natural potency of wilderness now became the property of the nation that had adopted it—in totem fashion—for symbolic purposes.[22] The national park system created places of refuge from development for fragile ecosystems, but it did not address the underlying issue of how to balance human, social, economic, and natural needs in the common spaces of the developing West. And it continued to rest on the fiction that the American land in these hallowed sanctuaries was static, unchanging, outside, and above the complex interplay of human and nonhuman nature—in short, the wilderness ideal.

Yellowstone's standing as a wilderness was achieved through a willed act of historical erasure. Setting aside public lands protected from development, the creation of national parks was also premised on the exclusion of any human presence (see chapter 7). Shoshone, Bannock, and Crow Indians had crisscrossed Yellowstone for centuries or lived—like the Sheep Eater Indians—in the park itself. Blackfeet, Coeur d'Alene, and Nez Perce Indians also passed through Yellowstone in their annual migrations. The area designated as parkland was marked throughout with the evidence of sustained human habitation. Native practices of annual burning had cleared paths, kept underbrush in control, and created the conditions in which certain plant species used by Indians could flourish. Obsidian was quarried from the rich deposits in the region and then traded over vast distances as part of an extended cultural network of influences. Moran's painting at least acknowledges this prior presence; yet in the same stroke, the masterly presentation of a landscape aesthetically reordered by the artist effects a form of symbolic dispossession, a dispossession confirmed by the implied narrative acted out in the two foreground figures.[23] The image of land transfer proved prophetic; the Crow Indians—demoralized by the depletion of game by commercial hunting, the near-extinction of buffalo, and the destruction by miners of their traditional grasses and other foodstuffs—ceded a good part of their lands within the park to the U.S. government in 1880. By the late 1880s, the park management had become adamant about excluding native hunters from the park, as recreational hunters complained of diminishing stocks and "wanton" destruction of game.[24] Throughout much of the nineteenth century, eastern Americans, far removed from the territorial tug-of-war in the West, had sentimentalized the plight of the "vanishing" Indian, whose fate was intertwined with an endangered wilderness. In an ironic reversal of these earlier sympathies, park constituents—hunters, tourists, and management—blamed the native presence in Yellowstone for the ruination of the wilderness. Moran's painting resolved these painful historical realities in a wishful image of the peaceful transfer of ownership and sovereignty from the Indian to the scientific, managerial, and commercial interests that played such a key role in the colonization of the West.[25]

How does the historian explain this reversal of the older alignment between native people and wilderness into the newer attitude—institutionalized in the National Park Service in 1916—that the centuries-long presence of western Indians was now incompatible with the concept of wilderness preservation? The Romantic concept of wilderness had included the native presence; indeed, Romanticism's idealization of the Indian as living in harmony with nature was part of a much longer history reaching back to the noble savage, as discussed in chapter 2, and forward into modernist primitivism, with its longing to escape the burdens of civilized existence. The later nineteenth-century version of wilderness, though, with its insistence on expunging any human history or presence, merely pursued the wilderness ideal—grounded in the radical segregation of human and natural histories—to its logical extreme.[26] As an Indian rights activist with an eye for irony decades later dryly observed, "There was no wilderness until the Whites arrived."[27]

Conclusion

Thomas Moran's *Grand Canyon of the Yellowstone* offers a grand summation of the western paradise, entirely receptive to the penetrating gaze and scientific expertise of the postwar nation-state. Yet if we look beyond the visual seductions of his painting, with its compelling knowledge of geology and its masterful management of complex detail, we encounter a persistent attitude toward wilderness as a space apart. From its origins in a form of adversarial Romanticism to its institutionalization in the philosophy of the National Park Service, the wilderness concept reveals very different histories in its movement across the nineteenth-century cultural landscape of the nation. Critical to any assessment of its cultural impact is the degree to which its various apologists have acknowledged that wilderness is a part of human history, not separate from it; that "humans" include the people indigenous to these regions; and that, as such, wilderness becomes a reality that is pliable and open to human intervention, rather than a static ideal impervious to time, history, and human desire for an intimate and productive commerce with the natural world.

One might well ask why American artists—with the possible exception of Cole—have been so quick, in general, to serve the moral evasions of a nation that seemed, as Herman Melville put it in *Moby-Dick*, "not so much bound to any haven ahead as rushing from all havens astern." But this is to hold them to a higher standard of accountability than we do ourselves, by assuming that it is possible to think beyond one's own historical horizon. And we cannot, in any case, look to representation for anything more than a temporary, and ineffectual, resolution of deep-seated historical dilemmas. Landscape paintings offered aesthetic solutions to problems whose origins were social—an imaginary stage on which to explore the role of nature in the nation's evolving identity. Yet the problems they engaged could only be fully addressed in the arena of democratic deliberation, debate, and choice. Americans would confront these difficult issues repeatedly in the coming century, as attitudes about nature's place in our national life fractured along regional, economic, and social lines.

Notes

1. Carolyn Merchant, *The Columbia Guide to American Environmental History* (New York: Columbia University Press, 2002), 71. However, see Colin Campbell, *The Romantic Ethic and the Spirit of Modern Consumerism* (Oxford: Blackwell, 1987); and Kenneth Myers, "On the Cultural Construction of Landscape Experience, Contact to 1830," in *American Iconology*, ed. David Miller (New Haven, Conn.: Yale University Press, 1993), 58–79, who argue for complicity between Romanticism and capitalist consumerism.

2. Carolyn Merchant, "Shades of Darkness: Race and Environmental History," *Environmental History* 8, no. 3 (July 2003): 381; Raymond Williams, "Ideas of Nature," in his *Problems in Materialism and Culture: Selected Essays* (London: NLB, 1980), 70–71.

3. Perry Miller, "The Romantic Dilemma in American Nationalism and the Concept of Nature," in his *Nature's Nation* (Cambridge, Mass.: Belknap Press of Harvard University Press, 1967), 198.

4. Letter from Cole to Luman Reed, dated March 2, 1836, reproduced in full in *American Paradise: The World of the Hudson River School* (New York: Metropolitan Museum of Art, 1987), 126.

5. The moving panorama was popularized in the 1840s, the decade after Cole's painting. See John Francis McDermott, *The Lost Panoramas of the Mississippi* (Chicago: University of Chicago Press, 1958).

6. See Matthew Baigell and Allen Kaufman, "Thomas Cole's *The Oxbow*: A Critique of American Civilization," *Arts Magazine* 55 (January 1981): 136–39.

7. Basil Hall, *Forty Etchings, from Sketches Made with the Camera Lucida, in North America in 1827 and 1828* (Edinburgh: Cadell and Co.; London: Simpkin & Marshall, and Moon, Boys, and Graves, 1830).

8. It is instructive to compare *First Harvest* with Durand's *Sunday Morning* of 1860, a classic example of the middle landscape.

9. These phrases are taken from Cole's letters from the mid-1830s, cited in Angela Miller, *Empire of the Eye* (Ithaca, N.Y.: Cornell University Press, 1996), 60–61.

10. George Perkins Marsh, *Man and Nature*, ed. David Lowenthal (Cambridge, Mass.: Belknap Press of Harvard University Press, 1965); see also David Lowenthal,

George Perkins Marsh: Prophet of Conservation (Seattle: University of Washington Press, 2000). Marsh's insights have been substantiated and expanded on by recent environmental history; see Andrew C. Isenberg, *The Destruction of the Bison: An Environmental History, 1750–1920* (New York: Cambridge University Press, 2000).

11. Marsh, *Man and Nature*, 279.

12. Marsh, *Man and Nature*, 35.

13. Francis Parkman, *The Oregon Trail: Sketches of Prairie and Rocky Mountain Life* (New York: Hart, 1977), 230, 24; see also passages on 98, 25.

14. See Roderick Nash, *Wilderness and the American Mind* (New Haven, Conn.: Yale University Press, 1981), 52–53, 60, 75–77, on James Fenimore Cooper's character Leatherstocking, whose attitudes toward nature and the pioneer mirror those of Parkman. Cooper, like Parkman, was no democrat.

15. This midcentury myth of the pioneer anticipates the full enunciation of the frontier myth with Frederick Jackson Turner's "Significance of the Frontier in American History," in *Rereading Frederick Jackson Turner: "The Significance of the Frontier in American History and Other Essays,"* ed. John Mack Faragher (New York: Holt, 1994), 31–60, delivered in 1893. The representation of the pioneer reveals that the ground for Turner's frontier thesis was already being prepared culturally by midcentury. Examples include Thomas Cole, *Home in the Woods* and *Hunter's Return*; Jasper Cropsey, *Retired Life* and *The Backwoods of America*; Asher B. Durand, *First Harvest in the Wilderness*; George Caleb Bingham, *Emigration of Boone across the Cumberland Gap*; and William Tyler Ranney, *Advice on the Prairie*, all dating from the late 1840s and 1850s. Cropsey's *Backwoods* is an American frontier version of Cole's *Wild State* from *The Course of Empire.*

16. The term "middle landscape" can be traced to Leo Marx, *The Machine in the Garden: Technology and the Pastoral Ideal in America* (New York: Oxford University Press, 1970), especially 121, 150, 226.

17. See Andrew Hemingway, *Landscape Imagery and Urban Culture in Early Nineteenth-Century Britain* (New York: Cambridge University Press, 1992); William Cronon, *Nature's Metropolis: Chicago and the Great West* (New York: Norton, 1991).

18. See the vast bibliography on native interaction with the North American environment in Carolyn Merchant, *The Columbia Guide to American Environmental History* (New York: Columbia University Press, 2002), 335–46.

19. See Robert H. Keller and Michael F. Turek, *American Indians and National Parks* (Tucson: University of Arizona Press, 1998); Mark David Spence, *Dispossessing the Wilderness: Indian Removal and the Making of the National Parks* (New York: Oxford University Press, 1999); and Philip Burnham, *Indian Country, God's Country: Native Americans and the National Parks* (Washington, D.C.: Island, 2000). Rebecca Solnit, *Savage Dreams: A Journey into the Landscape Wars of the American West* (Berkeley: University of California Press, 1999), dwells on the historical ironies of a sublime wilderness aesthetic built upon the often violent erasure of native histories.

20. Making a similar suggestion is Joni Louise Kinsey, *Thomas Moran and the Surveying of the American West* (Washington, D.C.: Smithsonian Institution Press, 1992), 44.

21. Landscape artists frequently benefited from commercial and industrial development; Frederic Church accompanied lumbermen to paint the wilderness of Maine, while

the railroad was ironically celebrated as a means through which Americans could have greater contact with nature.

22. See Myra Jehlen, "The American Landscape as Totem," *Prospects* 6 (1981): 17–36.

23. On Moran's reorganization of the actual landscape of the Yellowstone Valley, see Kinsey, *Thomas Moran*, 54–58. Kinsey also associates this aesthetic "management" of the site with the concurrent scientific survey of the region.

24. Spence, *Dispossessing the Wilderness*, 62; and Keller and Turek, *American Indians and National Parks*, 17–42.

25. See Alan Trachtenberg, *The Incorporation of America: Culture and Society in the Gilded Age* (New York: Hill and Wang, 1982), 11–37; also Joel Snyder, "Territorial Photography," in *Landscape and Power*, ed. W. J. T. Mitchell (Chicago: University of Chicago Press, 1994), 175–201; and Michael Bryson, *Visions of the Land: Science, Literature, and the American Environment from the Era of Exploration to the Age of Ecology* (Charlottesville: University Press of Virginia, 2002), 80–104.

26. See Merchant, "Shades of Darkness," 385–87, on the disconnect between wilderness preservation and social justice.

27. Quoted in John Cawelti, "The Frontier and the Native American," in *America as Art*, ed. Joshua Taylor (New York: Harper and Row, 1976), 137.

Seven

Wilderness Parks and Their Discontents

Benjamin Johnson

The turn of the twentieth century witnessed the transformation of wilderness as an idea into wilderness as practice: the creation of parks and other areas permanently set aside from settlement. By this point, a body of influential Americans had become convinced that America's abuse of nature had spiraled far out of control of such haphazard measures. Where were the enormous herds of buffalo that had once covered the Great Plains? Yellowstone, a few private ranches, and an isolated park in Canada housed the pathetic remnants of a population that had once numbered as much as 40 million. Flocks of passenger pigeons had once numbered in the billions, darkening the sky in flight and blanketing hundreds of miles of forest when at roost. Sports hunters and commercial hunters blazed away, convinced of nature's inexhaustibility. The last known passenger pigeon, Martha, died a lonely death in a Cincinnati zoo in 1914, more than a decade after the last verifiable sighting in the wild.[1] In parts of Wisconsin and Michigan, where majestic stands of red and white pines had once loomed, smoldering stumps stretched as far as the eye could see. Cattle and sheep trampled the meadows of even the most remote mountain valleys. Reflecting on such depressing developments, Theodore Roosevelt articulated a key conservationist sentiment when he concluded in 1897 that "[t]he frontier had come to an end; it had vanished."[2]

Wilderness preservation was but one part of the larger response of conservationists to what they understood as nothing less than a crisis of their civilization. While some conservationists emphasized the preservation of what they thought of as untouched nature, still more placed a premium on increasing the efficient and sustainable exploitation of nature. But most agreed in attributing America's profligate wastefulness to the stupidity and ignorance of common people, on the one hand, and the greed and power of the corporate world, on the other. Timber companies and market hunters would cut down every tree and kill every buffalo as long as it was profitable to do so, conservationists believed, and the masses were too ignorant and powerless to stop them. "There will be a period of indifference on the part of the rich, sleepy with wealth," John Muir wrote in 1901, "and of the toiling millions, sleepy with poverty, most of whom never saw a forest." Those who had seen a forest weren't much help, either, thought Muir. He damned most rural people as "[m]ere destroyers . . . tree-killers, wool and mutton men, spreading death and confusion in the fairest groves and gardens ever planted," and he expressed his hope that the government would "cast them out and make an end of them."[3] Franklin Hough, the first federal forest commissioner, was much more concerned with the material side of conservation than was Muir. But he was similarly troubled by the "unstable and transient class" of backwoods settlers who "are accustomed to regard the world around them as open for their use . . . in matters of pasturage for their stock, as well as forest products for their own supply."[4]

To preserve the remnants of American nature, whether to ensure continued supplies of natural resources or to protect wildernesses where Americans might find refuge, required that land be put under the control of the federal government, which would then hire expertly trained professional foresters and park rangers to administer them in the public interest. These professionals would be free from the rabble and from the manipulation of corporations alike. As Muir wrote of the "noble primeval forests," "God has cared for these trees, saved them from drought, disease, avalanches, and a thousand straining, leveling tempests and floods; but he cannot save them from fools—only Uncle Sam can do that."[5]

Uncle Sam found himself with more and more power over the nation's landscape. The 1864 federal transfer of Yosemite Valley to the state of California for use as a park, along with the 1872 establishment of Yellowstone National Park, were the founding acts of the national park system, which encompassed thirteen units when the 1916 establishment of the National Park Service created a federal agency devoted to park management. In 1891, the year before John Muir helped to found the Sierra Club, Congress authorized President Benjamin Harrison to create forest reserves. More than 13 million acres were soon designated as reserves, with President Grover Cleveland adding another 21 million in 1897. These reserves were transformed into national forests in 1907, to be managed by the U.S. Forest Service, now a division of the Department of Agriculture. The federal government, which had previously done everything in its power to transfer lands into private hands as quickly as possible, had committed itself to the permanent management of much of the nation's territory.

Exactly what Uncle Sam should do with his new national forests and parks would be the subject of acrimonious debate in the years to come. Advocates of wilderness preservation—that is, of managing large portions of these lands as reserves where Americans might still encounter nature in the raw—found themselves at odds with both those who lived in and near the new public lands and more materially minded conservationists. The conflict between wilderness advocates and locals is the focus of this chapter, while in the next chapter Char Miller discusses the conflicts and commonalities of those who considered themselves conservationists. Because a wide range of rural Americans continued to hunt, fish, gather, log, and farm in the new parks and forests, these conservation measures often criminalized their ways of making a living. Local people generally sought to maintain their subsistence practices in the face of efforts by public lands bureaucracies (and other locals) to prevent them from doing so.

This period's conflicts between wilderness advocates and rural Americans—which have only recently become the subject of sustained historical scholarship—reveal two important aspects of wilderness preservation in modern America. First, they show that an appreciation of wilderness competed not only against rapaciousness and greed (as its advocates understood matters), but also against different visions of the proper relationship between humans and nature, visions that left much greater room for permanent human use and occupancy. To oppose wilderness preservation was thus not necessarily to disdain or to ignore nature. Second, these conflicts show just how *modern* wilderness was. Although wilderness advocates sought refuge from the modern world in America's wildlands, and thus understood themselves as opponents of the artificiality and unhealthiness of life in the industrial era, in an important sense wilderness did not exist until they invented it. For an area to be uninhabited and far from the grind of daily life, as a wilderness by definition was, it had to be cleared of occupants and distanced from the lives of those who remained nearby. Thus, native groups were removed from newly formed national parks, and settlers nearby were denied access to hunting and trapping grounds. There was thus more wilderness at the end of the conservation era than at its dawn. As a result, rural Americans often found themselves forcibly alienated from the direct access to nature upon which they had relied, thereby pressured into the wage labor and urban life that made wilderness seem like such a necessary antidote to the ills of the modern world. For them, wilderness was not so much an antidote to modern life as one of its primary manifestations. This feeling left important legacies of rural hostility toward wilderness that still shape national environmental politics.

The Wilderness Ideal and Rural America

Although there were no federally designated wilderness areas until much later, from the late nineteenth century onward portions of the nation's new public lands system were managed to meet the expectations of wilderness advocates.

Indeed, wilderness tourism, whether in the form of hunting or camping, was an important focus of many of the early conservation bureaucracies. National parks, for example, were off-limits to commercial logging and mineral development. The 1916 legislation establishing the National Park Service instructed it "to conserve the scenery and the natural and historic objects and wildlife . . . by such means as will leave them unimpaired for the enjoyment of future generations."[6] Decades later, wilderness advocates would fault the park service for building too many roads, campgrounds, hotels, restaurants, and other amenities in the parks, but at the outset they praised the construction of the roads and rail lines that made them more accessible.

Even the more materialist goals of conservation could often be reconciled with the wilderness impulse. Big-game hunters like Teddy Roosevelt and his friends in the Boone and Crockett Club were drawn to the continent's wild places because there they could act out the manly virtues of the past that were now threatened by urban life. At the same time, they spent enough money in this endeavor to make their quest for the authentic a viable modern industry, parallel to extractive industries like the railroad and timber businesses. As a state forestry board noted in 1911:

> The sportsman, too, is a medium, together with the lumber companies and the railroad, through which the forests exert an economic influence upon the country. They furnish cover for the game which calls him out. In pursuit of that game he expends quantities of ammunition. He buys guns, tents, canoes, and endless other paraphernalia, in the production of which countless citizens gain their living. The ammunition bought from the retailer means renewed activities all along the line back to the charcoal burner.[7]

Rational economic development and wilderness preservation could occur simultaneously and harmoniously.

Permanent occupancy and use, however, were different matters. For its advocates, wilderness was and is a place untransformed by human beings, where those who need refuge from the modern world may find it. "Thousands of tired, nerve-shaken, over-civilized people are beginning to find out that going to the mountains is going home," proclaimed John Muir in 1901. "Awakening from the stupefying effects of the vice of over-industry and the deadly apathy of luxury, they are trying as best they can to mix and enrich their own little ongoings with those of Nature, and to get rid of dust and disease." Muir's contrast between urban-industrial life and the wilderness has remained a constant over the last century for wilderness supporters. The idea of wilderness as a place apart was decisively implanted in federal public lands policy with the 1964 Wilderness

Act and its famous definition of wilderness as an area "untrammeled by man," as discussed by Mark Harvey in chapter 11. Contemporary advocates of wilderness continue to insist on the importance of wilderness as a place beyond the human realm, even as they mix this idea with more recent ideas about nature. The Wilderness Society, for example, prominently displays the 1964 act's definition of wilderness on its Web site, next to its description of its mission to "[d]eliver to future generations an unspoiled legacy of wild places, with all the precious values they hold: Biological diversity; clean air and water; towering forests, rushing rivers, and sage-sweet, silent deserts."[8]

What places in the United States were "untrammeled by man"? In the early twentieth century, not many. John Muir's beloved Yosemite Valley—the major attraction of Yosemite National Park—was inhabited by the Indians who shared its name, as it had been for centuries before he wrote of its glories. The valley's open, parklike appearance owed as much to their regular setting of fires as to the timeless "Nature" that Muir invoked. Like Yosemite, most national parks had very recently been actively inhabited or regularly used by Indian peoples, as seen in Angela Miller's discussion of Yellowstone in the previous chapter. The lakes of Minnesota and Ontario where Sigurd Olson dipped his paddle may have been beyond "the steel and traffic of towns," but he traveled over portages worn by centuries of travel by Anishinaabe (or Ojibway or Chippewa), fur trappers, and backwoods hunters. Even the unlogged forest he gazed upon was already undergoing a profound transformation as the result of the suppression of fires and the accidental introduction of a disease that crippled the reproduction of the majestic white pine. Americans may go to the wilderness to escape their cities and jobs, but any area now classified as a wilderness under the provisions of the 1964 Wilderness Act has a rich human history, and often one that left a dramatic mark on the landscape.[9]

Americans continued to mark the landscape in the conservation era. By the twentieth century, the United States had become an industrial giant, producing nearly a third of the world's manufactured goods. The warm glow of electric lights replaced the smoky flicker of kerosene in many an urban home. The railroad network, the world's largest, made possible the sure and speedy delivery of goods ordered from such national retailers as Montgomery Ward and Sears, Roebuck, and Company. The telegraph allowed for nearly instantaneous communication across the continent. Workers performed the same tasks over and over again on the assembly lines that lay at the heart of the industrial economy, or on the "disassembly lines" that turned cows and hogs into steaks, sausage, tallow, and hides. Not even time itself was beyond the reach of the new industrial order; in 1883, railroad companies introduced the four time zones that Americans still use today. Watches and clocks replaced the sun, moon, and stars as ways by which Americans measured out their lives.

Despite this explosive transformation, however, many Americans continued to turn to the land itself for some or all of their sustenance. Farming remained the single largest occupation, and most family farms continued to grow crops and gardens and to keep animals for their own consumption and use. Those with too little productive land to make a living by farming could still plant gardens or clear a field for hay for a dairy cow or two, either on their own property or on areas used in common. Women and children could make valuable additions to their family's meals by gathering herbs and berries. Young boys often set snares or hunted for small game such as rabbits or edible birds. All family members could fish nearby streams, rivers, or lakes. Indeed, most rural men were opportunistic hunters year-round, and in some places they would go on extended fall hunting excursions with neighbors and family. Large game such as deer, moose, or wild hog could provide a large portion of a family's diet, whether it was the legal hunting season or not. "When we run short [of food] we just go out an' get another one . . . we salted down a lot of moose and fish," remembered August Stromberg of his boyhood in early twentieth-century Minnesota. Wood, whether from one's own land or not, could be used for building homes, fences, animal shelters, or watercraft. In colder climes, cutting firewood was a necessary—and free—ritual each fall.[10]

Some of the earth's bounty could also be traded for money. Wild ginseng could fetch a nice price from traders. Meat, particularly venison or moose, could be sold just as well as eaten, though one had to be careful not to run afoul of the game warden. Firewood wouldn't always sell for enough to make cutting and hauling it worth the effort, but merchants might pay a decent price for Christmas trees. Pigeons were good eating—as a nineteenth-century song went, "When I can shoot my rifle clear / At pigeons in the sky / I'll say good-by to pork and beans / And live on pigeon pie"—but that also made game dealers willing to pay good money for fresh birds that could be iced and shipped to fine restaurants in New York or Chicago. Fur trapping could also be a source of scarce cash. Even children could run small trap lines close to home, as August Stromberg did. "We kids, from the time we was ten years old we used to trap for our clothes. . . . Maybe you got three dollars for a mink or fifty cents for a weasel, but that was good money them days."[11]

Indians were among those who relied most heavily on the bounty of nature. By the 1880s, all previously independent Indian nations had been conquered by the United States and confined to reservations. Reservation life was supposed to make Indians self-supporting independent farmers (at the same time as the nation's farmers were finding their economic and political power eclipsed by the industrial economy). In the meantime, annual rations of food and clothing mandated by the treaties between Indian peoples and the U.S. government were supposed to be enough to support them. Conventional agriculture, however, enjoyed at best a limited success on reservations. It was alien to the traditions of many peoples and unsuited to the soils and climate of many reservations. Moreover, many Indians rejected it outright because it was explicitly intended to destroy their culture and to assimilate them into the mass of American society.

So hunting, fishing, and gathering continued to be particularly critical to American Indians, whether they took place on reservations, their private land, or territory that they considered their own despite the federal government's claims. The Yosemite Indians, for example, continued to derive most of their sustenance from trout, sweet clover, roots, acorns, pine nuts, fruits, and berries, until park regulations pressured them into working for wages. The Blackfeet, like many Indian peoples, sought to protect their rights to live off the land even when they were forced to cede much of it to the federal government. Tribal negotiator White Calf was successful in adding language to an 1895 treaty that preserved "the right to go upon any portion of the lands [ceded in the treaty] . . . to cut and remove timber . . . to hunt upon said lands and to fish in the streams thereof, so long as . . . they remain public lands of the United States." Blackfeet leaders believed that this language protected their usufruct rights in perpetuity, even when the 1910 founding of Glacier National Park put much of their former territory under the control of a different part of the federal government. (The park service, as they found out, saw things differently.)

Even Americans more fully incorporated into the industrial economy could find themselves in need of direct reliance on the nonhuman world. In the countryside or small towns, Americans deeply embedded in the market economy could find a safety net in nearby fields, forests, lakes, and streams. If drought or a collapse in grain prices ruined a year's harvest of wheat, or if a slowdown or strike led to unemployment, hunting or trapping could keep food on the table. (Many parks experienced sharp rises in poaching arrests during recessions; in Yellowstone, they quintupled during the 1907 slowdown.)[12]

Subsistence was important even for the workforces of some of the most modern and advanced industries. In Minnesota's Iron Range, for example, large mining companies produced much of the ore necessary for the nation's industrial growth, but at the same time their employees found the woods an indispensable resource during frequent strikes, lock-outs, and slowdowns. Indeed, the importance of the woods was one of the few things upon which both sides in the region's bloody labor battles could agree. "[T]he readers of the red-flag outfit have taken off their best clothes and have gone to the woods," sniffed the pro-company editor of the *Miner* in the aftermath of a violent 1907 strike, "in all probability to use some of the dynamite in blasting fish to fill their aching voids." Labor radical Andy Johnson was on the other side of the conflict, but remembered the same dynamic: "those old time pioneers who came here before and after the turn of the century . . . many of them settled out here in the woods because they were blackballed... because of their political activity on behalf of the working class."[13]

Those who continued to make some or all of their living from the land had their own ideas about nature and what constituted acceptable and unacceptable uses of it. These attitudes are of course hard to recover: such people generally did not write books and were more preoccupied with avoiding law enforcement officials than with leaving a cohesive record of their thoughts and actions. Oral histories, court transcripts, and diaries, however, suggest that many rural folk saw the natural world as a deeply human place, one bound up in the fabric of their daily lives. When describing the lakes and forests around them, residents of the Iron Range, for example, were more likely to orient themselves by referring to nearby homesteads, familiar trapping grounds, and places where important events in their own lives had occurred than by reference to the natural landmarks that wilderness advocates and outdoor enthusiasts used to navigate the same territory.[14]

Indian peoples, particularly those who remained in their ancestral homelands, also attached deep social significance to particular places. This was not only a matter of remembering particular places as favorite hunting grounds or sheltered campsites, but also of believing them to be spiritually important. To the Blackfeet, for example, the mountains of what would become Glacier National Park were part of Mistakis, the Backbone of the World. Some of their most powerful spirits resided there, and they believed that their ancestors had been given tobacco and horses at several of the lakes nestled high in the mountains.[15] Similarly, the Pueblo peoples of what is now called New Mexico believed that they emerged into the present world from a lake. Regular retreats to perform ceremonies and rituals near special lakes in the mountains around their settlements—including two lakes in the present-day Pecos Wilderness—thus kept them spiritually connected to the power of the lake of their emergence. Even the less powerful landscapes closer to home were—and still are—sacralized places. As historian William DeBuys writes, "[I]n the immediate vicinity of a pueblo... there may be any number of small shrines—here a pile of stones associated with hunting small game, there a rock outcrop where one seeks spiritual aid in so mundane an occupation as cutting the leather soles of moccasins." Countless Indian peoples also engaged in ritual journeys and hunts to mark critical times of passage in life, such as entering adulthood. Indian identity was thus deeply entwined with the landscape.[16]

Like all humans, country people could treat these landscapes in a reckless and even self-destructive way. The agrarian critics of the frontier discussed by Steven Stoll in chapter 4 and the conservationists of the early twentieth century were right, after all, to fear their society's voracious appetite for natural resources. A farmer might give in to the temptation to plow a steep hillside, whether pushed by falling grain prices or enticed by what a larger harvest might buy from the Sears catalog. Ranchers often ran more cattle than their pastures could sustainably support, and almost all eagerly participated in the campaign to exterminate the wolf. Indians, too, could be agents of environmental destruction. Driven not only by their own needs for food, clothing, and shelter, they participated in the fur and bison trades that trapped out beaver in many western streams and dramatically reduced buffalo herds on the plains.[17]

At the same time, however, rural people practiced their own form of conservation. Employing a set of beliefs that historian Karl Jacoby has termed "moral ecology," they drew clear distinctions between legitimate and illegitimate uses of nature. Providing food or other essentials such as heat for oneself or one's family was a right that justified even trespass on private property. The killing of animals and harvesting of products for cash sale, on the other hand, were viewed with far greater suspicion and were more likely to result in being turned in to the game warden or even in a kind of vigilante conservation. Rural people placed restrictions even on acceptable hunting. Most communities practiced some kind of "law of the woods," which in one rendering emphasized "never kill anything you do not need." These unwritten codes also included sanctions against specific forms of hunting. As one upstate New Yorker recalled of the 1890s, "[T]here was a universal code that deer should not be disturbed while 'yarding,' or in the breeding season, and this applied to game birds as well."[18]

Those who violated the tenets of moral ecology did so at their own risk. William Binkley, a resident of Jackson Hole, Wyoming, found this out the hard way. In the late 1890s, he was arrested for killing an elk outside of hunting season in order to give its meat to a sick and hungry neighbor. His neighbors paid the $100 fine. In 1906, however, after Binkley slaughtered hundreds of elk for the sole purpose of harvesting and selling their valuable tusks, a "citizens' committee" in the town ordered Binkley and his partners in crime to leave town or be "left dead . . . for the scavengers to devour."[19]

The Challenge of Conservation

The vigilante justice directed at environmental waywards like William Binkley came out of a complicated brew of rural uses and abuses of nature. On the one hand, Americans deeply and profligately abused nature, shooting out the passenger pigeon and nearly dispatching the bison in this period. On the other hand, there were environmentally conscious elements of American backwoods culture.

Conservationists had little interest in or use for this complex thicket of ideas and practices. A wide range of conservationists, from wilderness advocates like

John Muir to economics-minded managers like Gifford Pinchot, saw rural uses of nature as a deep threat to their own projects of reordering nature for the benefit of modern America. Backwoods economic activity was not productive and modern enough to appeal to the materialist side of conservation, but was disruptive and visible enough to undermine the notion that these landscapes were removed from the hustle and bustle of modern life. The new park and forest bureaucracies thus moved decisively to curtail and even ban such practices. Hunting, timber gathering, trapping, and other activities were outlawed from many areas altogether and subjected to strict regulation and stiff fees in others.

For some nearby residents, the establishment of a national park or forest did not necessarily mean much change in the fabric of daily life. State game laws establishing regular hunting seasons and banning unsporting if efficient means of taking game—typically, no fishing with nets, spears, or explosives and no hunting with dogs or lights—were already in place in most regions. And new regulations and managers did not always mean more effective or stringent enforcement. For other Americans, however, wilderness conservation fell like a thunderbolt. In the fall of 1898, for example, Havasupai Indians began leaving their village in Havasu canyon, a side canyon of the Grand Canyon, in order to hunt game and gather plants, as they had for as long as they could remember. The supervisor of the surrounding forest reserve was outraged, fearing that their presence would mar the scenic beauty of what was rapidly becoming a major tourist attraction. He ordered them to return to their village and to cease hunting and gathering on what was now government land. Since the forest entirely surrounded the Havasupai reservation, it had effectively become impossible for them to live off the land. They would have to think of some way to make up for the firewood and meat upon which they had always relied. "We got no meat. My family hungry," one Havasupai matter-of-factly informed his captor when arrested for poaching.[20]

Others all across the United States shared the Havasupais' complaints. The Blackfeet had larger hunting grounds still available to them, but faced similar obstacles if they crossed the line into the part of their traditional domain that had become Glacier National Park in northwestern Montana. The Yosemite were among those most subject to the rule of conservation bureaucracies, as they lived smack in the middle of the park. And many non-natives were directly affected as well. Hispanic settlers in northern New Mexico found themselves charged money—of which they had very little—by the forest service to run their cattle and to gather firewood on land that had belonged to their villages for generations. Miners in the small towns of northeastern Minnesota now dodged aggressive game wardens when they went out to hunt, and they faced arrest for timber trespass if they gathered wood from the abundant forests around them. In other places, the offense was as much spiritual as material. In 1906, residents of Taos Pueblo in northern New Mexico were enraged to find that the government had placed Blue Lake, high in the Sangre de Cristo Mountains above them, into the national forest system. One of the holiest sites in their religion and the destination of an annual pilgrimage, the lake was now open to all tourists who wanted to enjoy its cool waters.

The social gulf between conservationists and many rural Americans contributed to the park and forest officials' lack of concern for the impact of their policies. This was most starkly displayed when the people in question were Indians. The forest supervisor who banned the Havasupai, for example, fairly dripped with contempt when he explained:

> The Grand Cañon of the Colorado is becoming so renowned for its wonderful and extensive natural gorge scenery and for its open clean pine woods, that it should be preserved for the everlasting pleasure and instruction of our intelligent citizens as well as those of foreign countries. Henceforth, I deem it just and necessary to keep the wild and unappreciable [*sic*] Indian from off the Reserve and to protect the game.

Indeed, conservationist attitudes toward Indians during the formative period of wilderness preserves were much more hostile than either before or after. In the nineteenth century, prominent conservationist thinkers had lauded Indians' relations with nature. In the 1830s, for instance, artist George Catlin advocated setting up nature preserves in the West that would incorporate natives alongside wild animals and scenic areas. Later in the twentieth century, the invocation of Native Americans' supposedly harmonious relations with nature would become a major staple of environmentalist rhetoric. But living Indians who dared to interfere with wilderness preservation were another matter entirely. John Muir was thus typical in his dismissal of the Yosemite as "mostly ugly, and some of them altogether hideous" and in his celebration that Indians' removal from the wilderness meant that "[a]rrows, bullets, scalping-knives, need no longer be feared." Americans could now hear the "solemn call" of the wilderness.[21]

Although Indian peoples bore the brunt of particular animosity, they did not do so alone. Enormous migrations from Southern and Eastern Europe around the turn of the century deeply polarized America. Conservation bureaucrats, mostly native born and of Northern European descent, shared many widespread prejudices against immigrant ethnic communities. A Minnesota sportsman's diatribe against "pot hunters" characteristically singled out "foreigners" as the major culprits for improper and excessive hunting. "They do not hunt for the sport of it," he insisted, ". . . more game is killed by these people than by Americans who shoot during the closed season." William Hornaday's widely read book *Our Vanishing Wildlife* (1913) was nearly hysterical in its warning that

"Italians are spreading, spreading, spreading. If you are without them to-day, tomorrow they will be around you.... the bird-killing foreigner... will surely attack your wild life."[22]

In short, then, the early management of national forests and parks to serve the values of wilderness preservation had dramatic repercussions for many rural Americans. And yet the power of wilderness as an idea—the appeal that a return to untouched nature had for so many Americans—led conservationists to gloss over or even actively to hide these repercussions. Early conservation bureaucrats presented the areas under their control as natural and pristine even when they had abundant reason to know otherwise. In an interview with the magazine *Forest and Stream*, for example, the state game commissioner of Minnesota described the Superior National Forest as "an absolutely wild and unsettled country... there are no settlements or even settlers in the area, and nothing to attract them"—even as the nonexistent settlers harassed and shot at his game wardens. The forest service itself underscored the area's remoteness, even as it asserted its accessibility. "Fine camping sites are abundant, and the voyageur can always pitch his tent wherever night overtakes him—at places others have camped before, or perhaps where the ring of the woodsman's ax has never broken the forest silence."[23]

The hunters and campers traveling to such places—whether the woods of northern Minnesota, the Blackfeet country of Glacier National Park, the stark vastness of the Grand Canyon, or the splendor of Yosemite Valley—may have thought these places empty and silent wilderness refuges, just as Americans abroad would later think that tropical rainforests and other landscapes were untouched nature. But this was because the conservation movement had made wildernesses where before there had been none.

Consequences and Legacies

Some people, particularly Indians, were clearly removed from their traditional homelands and denied access to hunting grounds in the name of wilderness preservation. The full extent to which early conservation bureaucracies curtailed subsistence poaching and gathering, however, remains unclear. Successful poachers, after all, do not get caught nor brag of their exploits to the general public. It is clear, however, that the cost of breaking the law could be ruinous. At a time when bitter strikes were fought for a daily wage of $3, Minnesota's fines for possession of untagged venison ran around $25 plus court costs; hunting deer out of season was $100; and even netting fish was $10. When wardens sold confiscated goods, as they commonly did, moose meat and venison could fetch almost $20 per animal, mink fur $5, and more common furs $1 or $2. In Pennsylvania, the Italian immigrants who so often found themselves under the eye of the game warden generally earned under $2 a day in the early twentieth century. They could be fined $25 for carrying a gun "in the fields or in the forests or on the waters of this Commonwealth" without a license (which itself cost $10), and another $10 for any nongame bird in their possession. An additional $25 was added to the base fine if the illegal hunting took place on a Sunday. In 1905, the secretary of the state's game commission reported that an arrest of a poacher "seldom results in a penalty of less than $60 or $70 with [court] costs, sometimes very much more than this amount." The average fine was thus well over a month's wages.[24]

Not only were these fines and proceeds a significant blow to hunters, but the destruction, confiscation, or selling of canoes, traps, and guns also deprived rural Americans of the equipment necessary to engage in common subsistence activities. Wardens and park and forest rangers must have enjoyed at least some success in enforcing these regulations, or they would presumably not have been the targets of violence and even assassination as often as they were.

This remaking of rural America in the name of wilderness hit Indian peoples particularly hard. For them, wilderness preservation was a continuation of their conquest and dispossession by white America. Regaining access to traditional lands now encompassed by national parks or forests ranked high among the priorities of Indian peoples across the West. The process of dispossession in the name of wilderness took longer for some Indian groups than for others. Indians remained in Yosemite Park as actual residents for most of the twentieth century, for example. Even after park regulations made the Yosemite's traditional subsistence practices impossible to continue, they managed to earn cash working as guides, in hotels, as drivers of sightseeing wagons, and as maids and domestics. They lived in an environmental version of a company town, as park officials controlled nearly every aspect of their lives, including deciding who would be able to remain in the Valley. Officials punished theft and drunkenness with expulsion and tried to remove Indian residences from the view of campers and hotel customers. In the late 1920s, park superintendent Charles Thompson wanted to remove the Yosemite entirely, arguing that "they should have long since been banished from the Park" and that their ejection "would ease administration slightly; would eliminate the eyesore of the Indian village . . . and . . . would remove the final influence operating against a *pure status* for Yosemite." Even the Indians would benefit, as removal would "tend to break them up as a racial unit and, in time, to diffuse their blood with the great American mass."[25]

Thompson was too fearful of a backlash by the Yosemite and their allies to follow his own urges. Instead, he opted to build a new Indian village, intended to be more "traditional" looking, farther away from the most heavily visited portions of the valley. The park service took the opportunity to inform the Indians that their continued presence was a "privilege dependent upon proper deportment" and that anybody who "did not want to work reasonably steady, cannot get along with his neighbors, or in any way prove[s] to be a poor member of the Village... would have to go away and give up his house." The service allowed about fifty people, including the most skilled craftspeople and cooperative employees, to move into the new village, but banned some ordinary laborers and those who had clashed with the park's management. After World War II, housing was restricted to permanent government employees and their immediate families. As they retired or were dismissed, they were forced to pack up their belongings and leave the valley. The last resident, Jay Johnson, retired from the National Park Service and moved to Mariposa, California, in December 1996.[26]

Most Americans have never heard of Jay Johnson and his non-Indian counterparts, but nonetheless the conflicts in which they were involved left important legacies in the decades that followed the creation of the public lands system. Most rural communities still support subsistence hunting, even when it takes place out of season or on lands, such as federal wildernesses, where it is banned altogether. In places with substantial deer populations, many rural residents—even game wardens—hunt deer year-round. Other residents of such places will still report some violations of game laws to authorities: the indiscriminate slaughter of animals or illegal hunting by outside sport hunters is likely to result in a call to the warden. Local people often make similar distinctions between outsiders and insiders when it comes to their acceptance of entry regulations into parks, forests, and wilderness areas. Federal policy may treat wildernesses as places open to the national population for limited visitations, but for nearby residents they remain important community resources.[27]

Indian peoples have been particularly insistent on maintaining their access to places that the government has turned into wilderness. Struggles for control of land now in national forests and parks were major features of Indian politics for much of the twentieth century. The Blackfeet, for example, never stopped insisting on their right to travel and hunt in what is now Glacier National Park. In the 1930s, relations with the park service were so tense that historian Mark Spence described them as "a near state of war... with Blackfeet and rangers prepared to shoot and be shot upon at any given time." When the New Deal allowed for the formation of tribal governments, the Blackfeet became one of the first Indian groups to do so, soon filing lawsuits and petitions with the Indian Service. "Negative opinions of the park service," Spence concluded, "had become a central aspect of tribal policy and a fundamental expression of Blackfeet national identity."[28]

Indeed, Indian peoples have enjoyed some remarkable successes in recent decades in rolling back what they see as the excesses of wilderness preservation. The Taos Pueblo spent more than half of the twentieth century fighting for a return of Blue Lake to its control. Their efforts paid off: they won the right to return to the lake in 1927. They were forced to accept tourist access to the sacred site until 1970, when the federal government finally relented. In the 1970s, the Blackfeet tribe, though unsuccessful in advancing its larger claims on Glacier National Park, secured the waiver of entrance and camping fees for its members. Yosemite pressure resulted in a similar arrangement and in the creation of an Indian cultural museum in the valley that bears their name. The park service struck a remarkable deal with the Oglala Sioux in 1978, in which Badlands National Park was doubled in size but the Sioux "retained ownership of all reservation land within the new park boundaries." The Havasupai benefited from a similar arrangement with Grand Canyon National Park, regaining "traditional usage" rights and an outright expansion in the size of their reservation as the park, as well as the portion of the park under wilderness protection, grew. Today, dozens of tribes are pressing for more generous rights to access and use national parks and forests—and, in some cases, even for joint management.[29]

Wilderness preservation remains a point of contention among European Americans as well, although non-Indians have had less success in asserting local control over federal lands. Activist Lynn Laitala's 1997 speech to a local group opposed to further environmental restrictions in northeastern Minnesota serves as a good example of the dim view many rural Americans still take of wilderness advocacy, as well as the lens through which they see this history. She opened with an environmentalist elegy of a simpler time, when her family and community lived closer to nature:

> [O]n a bend of the Shagawa River, there used to be a row of boathouses standing on pilings. They had been built by the Finns to house the boats, which they had also built. . . . Our boathouse was the one closest to the rapids . . . and long after we got an aluminum boat and the three-horse Johnson motor, I would still row the length of the river to listen to the red-winged blackbirds sing from their cattail perches, creep up on turtles sunning themselves on rocks, and watch the great blue heron stepping high in the shallows. . . . my brother Gene and I would close the boathouse door behind us and sit in the dark, peering down into the river flowing gently under the slip. The water carried its own light which flickered and danced on the boathouse walls. . . . Minnows darted, big fish swam through more lazily. Frogs pumped along, waterbugs skated on the surface, and a couple of times, we saw a muskrat glide by.

Appreciation for this sort of reverie with nature, however, did not translate into a wilderness sensibility for Laitala. Rather than protecting this lifestyle from destruction at the hands of the modern world, for her wilderness preservation meant the transformation of this landscape into a playground for wealthy outsiders. Her speech abruptly shifted to a terse description of how a local canoe-outfitting business (owned by the Rom family) betrayed the history behind her family's boathouse:

> Then one day when I drove by, our boathouse was gone. One of the Rom kids had torn it down because he wanted the weathered boards to line his new business in Ely. He thought the rustic look would appeal to the tourists. All he had seen was a ramshackle building. He couldn't see the history it represented, or the visions it offered. . . . In their never-ending publicity campaign for the Boundary Waters [the large federal wilderness in the area], wilderness promoters depict their cause as saving the land. To me, it's been just another kind of development that wipes out a community for other types of use, in this case, up-scale tourism. We were assured . . . that the wilderness legislation would be good for Ely's economy. Well, it was good for some. Now there are four or five shops in town where you can buy $350 anoraks, but there isn't a store where you can buy school clothes.[30]

Contemporary wilderness advocates vehemently disagree with perspectives such as Laitala's. They speak of sacred places saved from the destructive impact of the modern economy, surely as threatening a force now as when the first national parks and forests were founded around a century ago. The depth of passion on both sides of this divide speaks to the ways in which wilderness still competes with other views of the proper relation of humans and nature and the extent to which it still shapes much of the modern American countryside.

Notes

1. Jennifer Price, *Flight Maps: Adventures with Nature in Modern America* (New York: Basic, 1999), 3.

2. Theodore Roosevelt, "The American Wilderness: Wilderness Hunters and Wilderness Game," originally published in *The Complete Works of Theodore Roosevelt*, as excerpted in *The Great New Wilderness Debate*, ed. J. Baird Callicott and Michael P. Nelson (Athens: University of Georgia Press, 1998), 70.

3. John Muir, *Our National Parks* (Boston: Houghton Mifflin, 1901), as excerpted in Callicott and Nelson, *The Great New Wilderness Debate*, 60–61.

4. As quoted in Karl Jacoby, *Crimes against Nature: Squatters, Poachers, Thieves, and the Hidden History of American Conservation* (Berkeley: University of California Press, 2001), 168.

5. Muir, *Our National Parks*, 61.

6. Hillory A. Tolson, *Laws Relating to the National Park Service, the National Parks and Monuments* (Washington, D.C.: Department of the Interior, 1933), 9–11.

7. Quoted in Benjamin Johnson, "Subsistence, Class, and Conservation at the Birth of Superior National Forest," *Environmental History* 4:1 (1999): 91.

8. Muir, *Our National Parks*, 47; Sigurd Olson, "Why Wilderness?" *American Forests* (September 1938): 395–96; United States of America Public Law 88–577, 88th Cong., September 3, 1964; http://www.wilderness.org/AboutUs/index.cfm?TopLevel=About (accessed January 2, 2004).

9. For a discussion of the Yosemite Indians and the impact of fires on the valley's vegetation, see Mark David Spence, *Dispossessing the Wilderness: Indian Removal and the Making of the National Parks* (New York: Oxford University Press, 1999), 102; for the transformation of northern Minnesota's forests in the early twentieth century, see Miron Heinselman, *The Boundary Waters Wilderness Ecosystem* (Minneapolis: University of Minnesota Press, 1996), 58–73, 109–11.

10. For descriptions of rural subsistence practices in this period, see Jacoby, *Crimes against Nature*, 23; Steven Hahn, "Hunting, Fishing, and Foraging: Common Rights and Class Relations in the Postbellum South," *Radical History Review* 26 (1982): 36–64; for Stromberg, see Johnson, "Subsistence, Class, and Conservation," 86.

11. Johnson, "Subsistence, Class, and Conservation," 86, for firewood and Christmas trees; for pigeons and the song, see Price, *Flight Maps*, 52, 35; Stromberg quoted in Johnson, "Subsistence, Class, and Conservation," 86.

12. Jacoby, *Crimes against Nature*, 137.

13. Johnson, "Subsistence, Class, and Conservation," 90, 92.

14. Johnson, "Subsistence, Class, and Conservation," 92.

15. Spence, *Dispossessing the Wilderness*, 73.

16. William DeBuys, *Enchantment and Exploitation: The Life and Hard Times of a New Mexico Mountain Range* (Albuquerque: University of New Mexico Press, 1985), 26; Jacoby, *Crimes against Nature*, 186.

17. For an extended discussion of Native Americans and the environment, see Shepherd Krech III, *The Ecological Indian* (New York: Norton, 1999).

18. Jacoby, *Crimes against Nature*, 3, 24.

19. Jacoby, *Crimes against Nature*, 139.

20. Jacoby, *Crimes against Nature*, 184.

21. Jacoby, *Crimes against Nature*, 175; Spence, *Dispossessing the Wilderness*, 109; Muir, *Our National Parks*, 55; Karl Jacoby, "Before the 'Ecological Indian': Conservationists and Indians at the Turn-of-the-Century," unpublished manuscript, 2002.

22. Johnson, "Subsistence, Class, and Conservation," 90; Louis Warren, *The Hunter's Game: Poachers and Conservationists in Twentieth-Century America* (New Haven, Conn.: Yale University Press, 1997), 26.

23. Johnson, "Subsistence, Class, and Conservation," 91–92.

24. Johnson, "Subsistence, Class, and Conservation," 89; Warren, *The Hunter's Game*, 28–29.

25. Spence, *Dispossessing the Wilderness*, 122–23.

26. Spence, *Dispossessing the Wilderness*, 122, 131.

27. Warren, *The Hunter's Game*, 104–5.

28. Spence, *Dispossessing the Wilderness*, 94, 98.

29. DeBuys, *Enchantment and Exploitation*; Spence, *Dispossessing the Wilderness*, 135.

30. Lynn Laitala, "CWCS Keynote," 18 May 1997, transcript in possession of the author.

Eight

A Sylvan Prospect
John Muir, Gifford Pinchot, and Early Twentieth-Century Conservationism

Char Miller

At the turn of the twentieth century, naturalist John Muir and forester Gifford Pinchot recognized that the American landscape—particularly its wooded wilds—was under assault. The industrial revolution's astonishing capacity to consume natural resources, for all of the beneficial economic growth it spurred, was devastating the nation's public domain. Like them, President Theodore Roosevelt was convinced that the citizenry must mend its ways if America the Beautiful was to retain a shred of its former grandeur. At the 1905 Forest Congress, which Pinchot organized to advocate the creation of the forest service he would head, Roosevelt raged against persistent land fraud on western public lands and western legislative resistance to the implementation of regulations that would control their sale, dispersal, and management. "You all know . . . the individual whose idea of developing the country is to cut every stick of timber off of it and then leave a barren desert for the homemaker who comes in after him," the president declared. "I ask, with all the intensity that I am capable, that the men of the West remember the sharp distinction that I have just drawn between the man who skins the land and the man who develops the country. I am going to work with, and only with, the man who develops the country. I am against the land skinner every time."[1]

Muir and Pinchot shared Roosevelt's antipathy for "land skinners," and in their differing attempts to alter how Americans writ large interacted with wilderness, they shaped the rhetoric and political tactics of succeeding generations of environmentalists. For these later environmentalists, Muir has been most closely associated with the idea of "preservationism," whose advocates argue for maintaining intact wilderness, free from human use; Pinchot, as a conservationist, was no less enamored of wildlands, but felt that human beings must utilize some of them for the resources necessary to human life. This difference in focus has been simplified around two key phrases: Muir's "temple of nature" has been set against Pinchot's maxim, "the greatest good of the greatest number in the long run." As a result, their intellectual legacies have provided ammunition to American culture's continuing debates over the meaning of wilderness and conservation. That's unfortunate, for these caricatures have masked the many ways in which these men worked together to articulate an environmental agenda new to the American polity and have denied their personal relationship its full complexity and their fragile political alliance its full depth.

Becoming John Muir

Founding president of the Sierra Club, Muir today is that venerable organization's poster prophet, the wise greybeard whose unshakable faith that an enveloping nature will restore our souls is blurbed on its annual desk calendars, drop quoted in slick coffee-table photo books, and cited in fundraising letters. But in his day, Muir stood as a different kind of archetype: this son of a first-generation Scottish immigrant family, with few advantages and a spotty education, rose to considerable prominence by dint of hard work and native intelligence, nature's Horatio Alger.

That's just as Muir intended it. Biographer Steven J. Holmes points out that one of Muir's "primary literary tactics—as well as a recurring rhetorical strategy within the environment—was to offer his life story as the embodiment . . . of a certain sort of personal experience of the natural world." So successful was this approach, so ubiquitous has Muir become, that he now "constitutes contemporary America's preeminent image of the 'Green Man,' an enduring mythic figure in Western culture."[2]

Muir's life story is gripping, revolving as it does around "a series of vivid images of his personal relationships with particular natural places." Born in Dunbar, Scotland, in 1838, the third child of Daniel and Ann Gilrye Muir, he grew up in a family of some comfort but whose success was masked by an austere home devoid of illustrations or music. Daniel Muir was, by all accounts, a religious zealot against whose restraints John would chafe. When he and his younger brother, without paternal permission, headed into the countryside, the sense of liberation was palpable. "[W]e were glorious, we were free," John Muir remembered. "[S]chool cares and scoldings, heart-thrashings and flesh thrashings alike, were forgotten in the fullness of Nature's glad wildness."[3]

The headstrong child developed into a rebellious youth, appropriately enough in the United States, where the family moved in 1849. This new terrain, in Wisconsin, which a young John fantasized would be a "wild, happy land," was not all that he had imagined. His father's grim insistence that John and his younger brothers bring the farmland to heel—Daniel Muir did none of the heavy lifting—led to countless hours of difficult plowing, stump removal, well digging, and animal husbandry. With little intellectual release, because his father had exiled such secular subjects as math, science, and literature from their frontier home, John had to sneak his reading under the covers at night, or down in the cellar before daybreak; his justly famed tinkering was also conducted during his scant off-hours. Clever, observant, and resourceful, at home in the intersection of the domestic and natural landscapes, Muir was ready to strike out on his own by his early twenties.[4]

He did so without paternal blessing, which no doubt was part of his point in leaving home, and he made his way to Madison, where he attended the state university for several years. After the cessation of the Civil War—which he had dodged by living in the Canadian wilderness—Muir worked in Indianapolis, where he proved adept as a machinist, quickly rising up the ranks. His peripatetic quest to find a fit between self and society, so emblematic of nineteenth-century American rootlessness, intensified when he was temporarily blinded in a workplace accident. If healed, he promised to undertake "the study of the inventions of God," and underscored his commitment as his sight returned by invoking a transcendent Christian ethos: "Now had I arisen from the grave. . . . I am alive!"[5]

The newborn prophet acted on his newfound principles, embarking on a thousand-mile odyssey through the war-ravaged American South. Enamored of its high country streams that coursed through "forest walls vine-draped and flowery as Eden," Muir began to reconsider once-central beliefs, and that led him, biographer Michael Cohen asserts, "consciously and categorically to deny, point by point, many nineteenth-century assumptions about God, Man, and Nature." Drawing on his reading of Ralph Waldo Emerson and Henry David Thoreau, he began to imagine a new environmental ethos that challenged claims based on Genesis that humans had absolute dominion over the natural world: "Why should man value himself more than a small part of the one great unit of creation?"[6]

Muir's question, and the biocentric consciousness it helped to develop, deepened during his years living in California's Sierra Mountains, a rough, glacier-forged landscape that he anointed the "Range of Light." He had moved west after his southern jaunt, convinced he might find the answers he was seeking in that golden state. The more he hiked through its stunning terrain, the more he felt compelled to capture its divinity, first in his notebooks, later in correspondence, then in newspaper articles, and finally in books. In each, he wrestled with how best to convey Yosemite's power to humble humanity, to realign its relationship to the larger world through which it moved. "Ink cannot tell the glow that lights me at this moment in turning to the mountains," he once confessed, yet his words hit the mark. Eastern periodicals published his accounts of life on the edge of civilization, of Nature awesome and terrible. His conception of modern man as the defiler of paradise, and his conviction that only when alone, in wilderness, could humanity approach holiness, melded with a cultural fascination for frontier wilds, which were then rapidly disappearing before the advance of the industrial revolution. "John of the Wilderness," and the jeremiads he preached, gained a growing body of adherents.

So did his demand that the federal government step in to avert environmental disaster in the Sierras and elsewhere. "Any fool can cut trees," he wrote. "They cannot run away."[7] But only government could save them. It was in this guise of celebrated citizen-advocate that Muir would meet the young Gifford Pinchot.

Eastern Forester

Pinchot's childhood could not have been more different from Muir's. Born in 1865 to James and Mary Eno Pinchot, he was the eldest child of one of New York City's elite mercantile families. His maternal grandfather, Amos Eno, had amassed a fortune through urban land speculation, and his father did so through the distribution of domestic and commercial furnishings, retiring in his forties.[8]

Had his grandfather had his way, Gifford never would have become a professional forester and public servant. But James and Mary Pinchot were determined that their first son pursue a higher calling than the pursuit of mammon. One reason for their determination was tied to a source of the Pinchot family's financial well-being—lumbering—and to James's reactions to its impact on the land.

James Pinchot's father, Cyril C. D. Pinchot, had bought forested lands in eastern Pennsylvania, in and around Milford, where the family had settled. He then clear-cut them, shipped log rafts downstream to market, sold the open land to farmers, and plowed the financial returns into another timber stand. The environmental consequences of this cycle were considerable. Unregulated by anything other than market demand, lumber entrepreneurs like Pinchot ripped through old-growth forests, leaving behind denuded hills, eroded terrain, and silted rivers. The scars in the landscape only deepened when, at midcentury, to feed the railroads' gargantuan appetite for wood, they employed more technologically advanced and efficient means to cut, transport, and mill timber.[9]

It was not until the late nineteenth century that James Pinchot and his contemporaries began to question the received wisdom about the unregulated human use of nature. Pinchot's emerging distaste for the rapid deforestation of his corner of the continent was first expressed in his growing preference for the seemingly bucolic English countryside, with its neat fences and greensward, which stood in dramatic contrast to the blighted environment and dust-filled air of his hometown.[10]

Aware that those stumps that so marred the rolling hills of northeastern Pennsylvania were a direct consequence of his family's lumbering activities, James Pinchot admitted that their economic gain had come with an environmental cost. Lumbering, once "the great business of the upper Delaware valley," had helped to transform the region: "[t]he [passenger] pigeons are extinct, the smaller brooks where we fished are often dry, venison and bear meat are luxuries, and the forest from which they all sprung has largely disappeared." Springing from this was his recognition of the importance of forestry, symbolized in his early membership in the American Forestry Association and in his family's later endowing of the Yale School of Forestry.[11]

James and Mary Pinchot also offered up their eldest child, Gifford, as compensation for past misdeeds. It was only after "seeing forestry practiced in France that a solution came to my mind," James wrote, and that entailed advising "my son Gifford to make forestry his profession, which he did." That there was no forestry profession in the United States at that time made this advice groundbreaking. It also meant that Gifford would have the field to himself, no small allure for an ambitious young man. Enticing too was the possibility that there was virtue in doing good that went beyond the mere getting of money. "No man can make his life what it ought to be by living it merely on a business basis," Gifford wrote in *The Fight for Conservation* (1910). "There are things higher than Business."[12]

After graduation from Yale College in 1889, Gifford Pinchot traveled to Europe, enrolling at L'Ecole Nationale Forestière at Nancy, France. On his extended tramps through French woods, he gained his "first concrete understanding of the forest as a crop," and how forestry could maintain "a permanent population of trained men" and "permanent forest industries, supported and guaranteed by a fixed and annual supply of trees ready for the axe."[13]

Forestry's emphasis on rational planning captivated the young man whose father had been so dismayed by the untidiness of the American countryside. The son in turn was struck by the systematic manner in which the French forests "were divided at regular intervals by perfectly straight paths and roads at right angles to each other." Elements of this startling demonstration of scientific management, of nature controlled through human stricture, in time would feed easily into the progressive ethos with which Pinchot would be so closely associated in the United States.[14]

But Pinchot also infused European scientific forestry with a progressive concern for social amelioration. As historian Clayton R. Koppes has argued, equity was a central ideal of many Progressive Era conservationists, Pinchot among them. To these reformers, "efficiency was but a means to the greater goal of equity," as reflected in their assumption that natural resources "belonged to all the people and should be retained in public control to prevent their concentration in the hands of the few." Only in that way could the "benefits of resource development be distributed widely and fairly," or as Pinchot was wont to trumpet: "For whose benefit shall [natural resources] be conserved—for the benefit of the many, or for the use and profit of the few?"[15]

Muir and Pinchot

What brought Muir and Pinchot together was a combination of Muir's growing acclaim and Pinchot family ambition. The setting of their first encounter in June 1893 was not in the sunlit valleys of Yosemite, in which Muir cultivated the friendship and support of so many of the luminaries of the early conservation movement, including Theodore Roosevelt, but in New York City, at a dinner party in the Pinchots' elegant Gramercy Park home where, Muir wrote to his wife, he was entertained in "grand style."[16]

The senior Pinchots were lavish entertainers and collected intriguing dinner guests as avidly as any big-game hunter stalked trophies. When they learned that the celebrated John Muir was in town prior to sailing for Europe, they immediately invited him to dinner, and that evening he played the role of Wild West raconteur, albeit in full dinner dress. Wilderness advocacy might have made little headway in such civilized environs.[17]

Then again, maybe it did, for Muir and young Pinchot became close, leading to subsequent invitations to Muir to spend evenings at the Pinchot home in New York and at Grey Towers, the family's estate overlooking the Delaware River in Milford, Pennsylvania, which Muir called a "cottage in the hills." Pinchot's parents hoped that Gifford's interactions with the great spokesman for America's forests would boost their twenty-eight-year-old son's fledgling forestry career.[18]

They were right. Early correspondence between the two men suggests that the fifty-five-year-old Muir gladly took up the role of mentor. "Nothing in all my trip gave me greater pleasure than finding you a Young Man devoting yourself to the study of World Forestry amid the whirl of commerce," Muir wrote after returning from Europe. That Pinchot's fix on forestry was based in part on a rejection of commercial enterprise—and was thus in concert with Muir's mountain thinking—was accurate to a point. Pinchot's grandfather Eno could not understand his energetic grandson's choice of profession, but he failed to change his grandson's mind. That did not mean, as Muir implied, that Gifford had rejected the commercial world. Becoming a public servant, for which he would earn little, was made possible by the very financial success he had spurned.[19]

Pinchot also treasured his mentor's advice to learn forestry through living in American forests. Muir had called the experience "getting rich," an ironic redefinition of the phrase, and Pinchot worked hard to collect this form of wealth. He wrote in the spring of 1894 that he had been "trying to live up" to Muir's expectations. That May, while working on a forestry plan for George W. Vanderbilt's Biltmore estate, he put words to action. "In a very small way I have tried your plan of going alone, and was off for four days by myself. They were as pleasant days as I have ever passed in the woods, and I am only waiting for the chance to do more . . . [for] I am perfectly satisfied that I can learn more and get more out of the woods than when there is anyone else along." Living up to his mentor's code did not mean that Pinchot thought he could ever eclipse Muir, however. "I am afraid that I shall never be able to do the amount of hard work that you have done, or get along on such slender rations," Pinchot acknowledged, but he hoped that by following Muir's path he might "be able to get more into the life of the forest than I have ever done before."[20]

Muir applauded Pinchot's efforts. "You are choosing the right way into the woods," he responded. He urged the younger man to press on with his work in the woods. "Go ahead. Yours must be not merely a successful but glorious life." Indeed, he challenged Pinchot to give his ambition free rein: "Radiate radiate radiate far and wide as the lines of latitude and longitude on a globe," he wrote in Whitmanesque exultation. "You have a grand future and a grand present." That exultation he would no doubt live to regret when in the early twentieth century Pinchot radiated well beyond Muir's orbit, and thus out of his control. But that trajectory lay in the future. For now, Pinchot's career seemed in tandem with Muir's vision. The older man's only regret was "that I cannot join you on your walks."[21]

That regret lasted until the summer and fall of 1896, when, as members of the National Forest Commission, the two men shared many a hike. The secretary of the interior, Hoke Smith, had asked the National Academy of Sciences to create a commission to study the nation's forest reserves and to prepare a report on their future purpose, resources, and management. It was in this context that Muir and Pinchot traveled together, but their shared physical jaunts did not necessarily mean that they thought alike.

For one, they conceived of their work for the commission in quite different terms, reflecting the differing calculations each man made about his ability to contribute to the group's deliberations. Not an official member, Muir traveled as an observer to retain his independent voice even while influencing the commission's findings. As the youngest member, younger by a full generation in some cases, Pinchot believed that becoming the commission's secretary, and thus a member of its executive committee, would give him access to the centers of power in which the final report would be written.[22]

These behavioral variations offer insight into the two men's political perspectives and personal styles. Muir was uncomfortable as a joiner, while Pinchot would prove to be the consummate organization man, contrasting traits that lay at the heart of their later ideological disputes. This was the period when they began to sense that their interpretations of conservation might not always be the same, that when they looked at trees, they did so with different eyes. Muir confirmed this indirectly when, in a letter to his wife, he commented that of all the members of the commission, only its head, Charles Sprague Sargent, Harvard botanist and director of the Arnold Arboretum, saw trees as he did; Pinchot, by contrast, lashed out at Sargent's deficiencies, believing that the Harvard professor "couldn't see the forest for the trees."[23]

It is tempting to imbue this disagreement with all of the significance of those that would come, to suggest that the two men had never been kindred spirits, but that would oversimplify a complicated human relationship. As their letters and diary entries indicate, they delighted in each other's company during the commission's tour, so much so that when Muir decided to take leave of it in July for a quick trip to Alaska, he invited Pinchot along. The younger man accepted with alacrity but was unable to go at the last moment due to a conflict with an inspection of Montana's Bitterroot Mountains. "You will know, without any words from me, how sorry I am that matters turned out in this way," Pinchot wrote. "I had already written home that I was going with you, and I know how sorry my people will be when I tell them . . . that the plan is changed."[24]

Their friendship deepened that summer in response to their shared concern over the deplorable state of the American forests and in lighter ways as well, most visibly in the adolescent bravado with which they displayed their common enthusiasms. One evening in Oregon, when their colleagues chose to sleep in cabins, Muir and Pinchot bedded down under the stars in an "alfalfa mow." Several days later Muir, Pinchot, Sargent, and a fourth member of the commission, Arnold Hague of the U.S. Geological Survey, rowed across Crater Lake to inspect the island that rises at its center, only to be forced back to shore when a violent thunderstorm swamped the overloaded boat. When the sodden crew regained land, Muir and Pinchot broke away from the others, scampered up a steep hillside, and reached a rocky ledge about one hundred feet above the lake, where they built a fire to dry out their clothing. That night, Muir noted in his diary, Pinchot alone slept outdoors in a driving rain, an act that could only have endeared him to his mentor. "That was the sort of behavior," biographer Cohen notes dryly, "which would go a long way toward making Muir forget other indiscretions." For the rest of the 1890s, they constantly made reference to their excursion, drawing upon its reservoir of good will and intense feeling to sustain cordial relations; camping bred a special kind of male bonding.[25]

This connection was easy to maintain when differences of opinion over forest policy did not loom large. In July 1897, for example, Muir wrote Pinchot to congratulate him on his appointment as special forest agent, urging him to do "grand work for Yourself and for all of us." Muir understood that by accepting this "most responsible position especially under present conditions," Pinchot could effectively preserve the size and character of the troubled reserves. "In running the new boundaries of the new reservations no doubt small changes should be made," but for "every acre you cut off, fail not I charge you to add a hundred or a thousand."[26]

The reciprocal cordiality of their correspondence, and the shared love of the natural world they reveal, are all the more impressive when one realizes that these two men had reached points in their thinking at which such reciprocity was increasingly difficult to manage. The bonds of language and affection could only stretch so far, and by the late 1890s the limit had been reached. By then, a new stage in their private correspondence and public relationship emerged in which the previous roles of mentor and student were no longer applicable, and a more discordant tone set in that would characterize their interactions until Muir died in 1914.

The potential for political disagreement was first manifest during the discussions of the National Forest Commission's final report. In the group's official deliberations and private conversations, it became clear that its members were divided roughly into two camps. Sargent, Muir, Alexander Agassiz of Harvard University, and Henry Abbott of the Army Engineers believed that the only way to preserve the reserves was to close them to development and that the best way to keep them inviolate was to deploy the U.S. Army. Pinchot and Arnold Hague disagreed sharply. The forests were to be used, they argued, not closed off; the most effective force to ensure their protection and regulated use was the creation of a professional, nonmilitarized, civil agency—a forest service—along the lines of those Pinchot had examined while studying forestry in Europe in 1890.

This dispute threatened the composition of the report, as Pinchot acknowledged to Muir just prior to the final series of meetings: "I am somewhat anxious to know just how the cat will jump. It is a rather critical time." The cat did not jump as Pinchot hoped. The commission voted in favor of Sargent's proposals, urging President Grover Cleveland to set aside vast tracts of public lands that would be closed to all development, except for mining and lumbering; the army would police these terrains. The preservationists appeared to have won the first round.[27] The division did more than just forever set Sargent and Pinchot apart; it foreshadowed the impending split between Pinchot and Muir.[28]

Why was the break between them not already overt? Part of the answer was that they liked one another, and their affection trumped the emerging philosophical disagreements. Moreover, it was not clear exactly how deep those differences ran. Muir had not yet resolved the key question of whether preservation and conservation were incompatible. His essays in the influential periodical *Century*, published during the period of the forest commission's work, indicated that he embraced Pinchot's idea that national forests should be preserved and *used*: "It is impossible in the nature of things to stop at preservation," the one-time sawyer declared. Forests, "like perennial fountains, may be made to yield a sure harvest of timber, while at the same time all their far-reaching uses may be maintained unimpaired." That balancing act was still evident in his 1897 essay, "The American Forests," in which he directly praised his young ally's work.[29]

Their alliance was strategic. For Muir, the principles of scientific forest management were a considerable advance over the slash-and-burn tactics that generations of Americans had employed in their conquest of the continent. Forestry seemed to promise the survival of trees, and thus of wilderness, and he happily joined with Pinchot, one of its chief advocates. Muir's support of forestry was equally crucial for Pinchot. Without Muir's eloquent voice and sharp pen rallying on behalf of forests and forestry, the public's interest in them would not have been as great. Without that interest, Pinchot knew better than most, there would be no legislation supporting the reserves or for the establishment of a national forest service; if there were no forest service, there would be no career in federal forestry for Gifford Pinchot.

Creating a Bureaucracy

This web of mutuality would unravel under the pressure of shifting circumstances. In 1898, Pinchot was appointed the fourth head of the Division (later Bureau) of Forestry in the Department of Agriculture. He would prove an able and aggressive bureaucrat, ever increasing his division's budget and rapidly expanding its workforce, all the while generating reams of positive publicity for his profession, building it and his division's successor, the forest service, into one of the most potent bureaucracies in American political culture. He would have little use, thereafter, for the voluntarism that had characterized the forest commission's activities, or for the preservationist visions that came to dominate its proceedings; by extension, his need for John Muir decreased, too. Now an insider, Pinchot gained authority through the political networks he constructed in Washington and nationwide and through the technological and managerial solutions he brought to bear on environmental matters. In this context, only the language of utilitarian conservation was spoken.[30]

Pinchot joined with other like-minded federal scientists to establish within the Washington bureaucracies a powerful force in support of what would come to be called the conservation movement, and which at the time was broad enough to include those, such as Muir, who advocated preservationism. Their collective attempt to manage a wide range of natural resources and to address a broad array of attendant social problems received great impetus when Theodore Roosevelt entered the White House following William McKinley's assassination in 1901. By pursuing legislative initiatives to expand federal control over public lands, waterways, and irrigation projects, Roosevelt firmly rooted conservation in public policy, laying the groundwork for environmental management on a national scale, a prelude to the New Deal.[31]

Theodore Roosevelt's commitment to conservation was profound. It was embodied in the coterie of bright, idealistic federal scientists and bureaucrats the president tapped to advance conservationism throughout his tenure in the White House. There were, as well, innumerable on-the-ground successes. With considerable public support and impetus, Roosevelt signed legislation establishing 150 national forests, and in 1905 he created the U.S.D.A. Forest Service, which Pinchot headed, to regulate their use; he created five new national parks, including Oregon's Crater Lake (1902) and Colorado's Mesa Verde (1906); with the power granted him under the Antiquities Act (1906), Roosevelt named the Grand Canyon a national monument and then cast that protection over seventeen other landmarks, ranging from Wyoming's Devils Tower to Washington's Mount Olympia. When at the 1912 Progressive National Convention Roosevelt asserted, in a speech aptly titled "Confession of Faith," that conservation was the preeminent issue of his time, he had practiced what he preached.

Roosevelt's presidency was a heady time for conservation activists. But its federal orientation tended to centralize authority in Washington and to increase reliance upon a corps of experts ensconced in distant bureaucracies; local autonomy appeared to have waned. Although Pinchot and his fellow reformers ardently believed that the national regulation of resources would ensure their equitable distribution, the reverse could also be true. Corporate control of land, water, and forests increased during this age of reform, as monopoly clashed with equity.[32]

The tensions accelerated when President William H. Taft took office in March 1909. Pinchot came to believe that Roosevelt's successor was undermining progressive environmental regulations, and he began to challenge administration policies, including exposing what appeared to be illegal coal leases on public lands. When President Taft fired Pinchot in 1910 for insubordination, his dismissal, although predicated on bureaucratic in-fighting and personal animosity, also revolved around differing interpretations of who should control national resources, and to what end. In firing the forester, and thus discarding Roosevelt's energetic conservationism, Taft sided, Pinchot wrote, with "every predatory interest seeking to gobble up natural resources or otherwise oppress the people." The president had become the "accomplice and the refuge of land grabbers, water-power grabbers, grabbers of timber and oil—all the swarm of big and little thieves and near-thieves" who sought to steal resources that "should have been conserved in the public interest." Its claims must remain paramount, Pinchot concluded, for that was the only way to make "the people strong and well, able and wise" and to build a nation "with equal opportunity for all and special privilege for none."[33]

The Struggle for Hetch Hetchy

Pinchot's progressive understanding of the dangers of "big and little thieves" would also bring about his famed fight with Muir over the Hetch Hetchy Valley in Yosemite National Park. This spectacular landscape, carved out of the Sierras by glaciers and the Tuolumne River, had become part of the park in 1890 and was designated a "wilderness preserve," a status for which *Century* editor Robert Underwood Johnson, John Muir, and others had long fought. But as early as the 1880s, San Francisco's water board and politicians had discussed the possibility of constructing a dam at the narrow end of the valley, creating a much-needed reservoir for the City by the Bay. Those plans were revived early in the twentieth century, and in 1903 and 1905 San Francisco applied to Ethan A. Hitchcock, the secretary of the interior, under whose jurisdiction the valley lay, for permission to build the dam. Hitchcock denied these early requests, indicating that they violated the spirit of the national park, but not before requesting that Pinchot examine the question. He obliged, assuring the secretary that the dam would not "injure the National Park or detract from its beauties or natural grandeur," an assurance that amazed Muir. "I cannot believe Pinchot, if he really knows the valley, has made any such statements," he wrote to Johnson, "for it would be just the same thing as saying that flooding Yosemite would do it no harm."[34]

But Pinchot had so reported to Hitchcock, as Muir learned after writing directly to Pinchot, seeking confirmation of his views. The forester noted that for him "the extreme desirability of preserving the Hetch Hetchy in its original beauty" must be weighed against the water needs of "a great group of communities" in the Bay Area. The people's public health in this instance took precedence over wilderness preservation. Muir challenged the basis of Pinchot's approach. Ignore the "benevolent out cry for pure water for the dear people," he replied, for the "scheme for securing these water rights is as full of graft as any of the lumber companies to obtain big blocks of the best timber lands." Besides, if the object were simply water, it "can be obtained below Hetch Hetchy, tho' at a greater cost. The idea that San Francisco must go dry unless Hetch Hetchy Yosemite is drowned is ridiculous."[35]

The debate intensified in April 1906, when a massive earthquake destroyed San Francisco, bursting water and gas pipes and setting off a fire that incinerated much of the city's housing stock and industrial base. Capitalizing on the wave of national sympathy for its plight, the city reapplied for permission to dam Hetch Hetchy. Pinchot was at the ready: "I was very glad to learn from your letter . . . that the earthquake had damaged neither your activity nor your courage," he wrote to city engineer Marsden Manson. "I hope sincerely that in the regeneration of San Francisco its people may be able to make provision for a water supply from the Yosemite National Park. I will stand by to render any assistance which lies in my power." Pinchot's assistance, especially when combined with the support of the new secretary of the interior, James Garfield, a close friend, produced the desired result: San Francisco received administrative approval to proceed with its plans for the valley.[36]

Although Congress turned back this effort, due to a storm of protest that Muir and Robert Underwood Johnson orchestrated, the issue did not disappear; San Francisco still needed a secure source of potable water. In 1907, Muir and Pinchot met in California to discuss Hetch Hetchy, during which Pinchot admitted that he had never seen the valley and, according to Muir, "seemed surprised to learn how important a part of the Yosemite Park the Hetch Hetchy really is." Pinchot suggested that Muir write to Secretary Garfield and request that he "keep the matter open until [the Sierra Club] could be heard." In September, Muir fired off an extended description of the valley to Garfield. Less than a month later, Pinchot wrote to the president that, although he fully sympathized with Muir and Johnson's position, "I believe that the highest possible use which could be made of [Hetch Hetchy] would be to supply pure water to a great center of population."[37]

The issue remained unresolved until 1913, and then only after Pinchot and Roosevelt were long out of office. President Woodrow Wilson and his secretary of the interior, Franklin Lane, former city attorney of San Francisco, signed off on Hetch Hetchy becoming a reservoir after hearings at which Pinchot testified. Simply put, his position was that the public welfare was of preeminent importance: "Injury to Hetch Hetchy by substituting a lake for the present swampy shore of the valley . . . is altogether unimportant when compared with the benefits to be derived from its use as a reservoir." To make this claim, those benefits must be widespread, and he believed that they were, and it was on this point that he sought to turn the political tables on Muir, who set aesthetic beauty and the preservation of wilderness before human need. Something was wrong with keeping the valley "untouched for the benefit of the very small number of comparatively well to do to whom it will be accessible," he declared. "The intermittent esthetic enjoyment of less than one per cent is being balanced against the daily comfort and welfare of 99 per cent"—and the scales necessarily tilted in favor of the masses, whose need for a publicly controlled water supply, safe from earthquakes and monopolists, was essential. The dam at Hetch Hetchy was a matter of equity.[38]

The masses be damned: Muir and his supporters were certain that the San Francisco project was not a democratic initiative, but simply the product of former San Francisco mayor James D. Phelan's "political ambition," or so Muir complained to Johnson. It also dovetailed with the perceived needs of powerful figures such as sugar magnate Claus Spreckles: Muir alleged that the dam's water power would run the city streetcars that Spreckles longed to own. That Pinchot could cozy up to these San Francisco capitalists was perverse, Muir asserted, shaking his head at what might have been: "I'm sorry to see poor Pinchot running amuck after doing so much good hopeful work—from sound conservation going pell-mell to destruction on the wings of crazy inordinate ambition."[39]

Twinned Legacy

Although Muir and Pinchot fought over their differing conceptions of wilderness, their internecine struggle did not wound the early conservation movement but rather was essential to its intellectual development and political success. Their ideological differences, and the psychological context in which those differences emerged, were critical to the creation of a generative dialogue between preservationists and conservationists and with the publics they expected to influence. Without such tension, the idea of conservation would not have emerged as one of the most important of Progressive Era credos.

Similar debates between radicals and reformers have shaped other movements committed to the transformation of American life. Antebellum abolitionists struggled with one another almost as much as they did with those steadfastly opposed to any form of slave manumission, disputes that gave greater focus to their shared cause. More than a century later, the Reverend Martin Luther King, Jr., and Malcolm X asserted contrasting visions for how best to destroy southern segregation and racial inequality, but their clashing perspectives served to widen the reach of the other's demands. These controversies, at once idealistic, intemperate, and rhetorical, often have served a larger purpose, as each segment gained by the other's presence. Radicals, for instance, can make moderates appear more conservative to those who fear reform, and as a result the moderates can often secure greater success. Pushed by the logic of confrontational politics, moderates often are compelled to adopt elements of the radical agenda to maintain their standing in a particular movement.

This is not to suggest that social change is inevitable, that history is inherently progressive, but neither should we miss the import of intramovement dynamics. John Muir and Gifford Pinchot succeeded, with the timely aid of President Theodore Roosevelt, because they fed off one another's ambitions and ideals, a dynamism that made for a much more potent conservation movement in the early twentieth century. This claim is underscored by the two men's institutional legacies and wildlands heritage. Out of their intense debate, at once personal and political, emerged the national park and national forest systems that today envelop hundreds of millions of acres; without their creation, it would have been immeasurably more difficult later to enact the Wilderness Act (1964) and to create designated wilderness areas, most of which have been drawn from the national forest inventory. These two men's enduring impact is revealed as well in the degree to which subsequent generations of American environmentalists have employed the rhetoric and tactics they devised during the Progressive Era. John Muir and Gifford Pinchot imagined the world we live within.

Notes

This chapter is revised from my *Gifford Pinchot and the Making of Modern Environmentalism* (Washington, D.C.: Island, 2001) and "The Greening of Gifford Pinchot," *Environmental History Review* 16 (Fall 1992): 1–20; I am grateful to journal editor Adam Rome for permission to republish.

1. Theodore Roosevelt, "The Forest in the Life of the Nation," *Proceedings of the American Forest Congress* (Washington, D.C.: Suter, 1905), 4–5, 10–11.
2. Steven J. Holmes, *The Young John Muir: An Environmental Biography* (Madison: University of Wisconsin Press, 1999), 3–4.
3. Holmes, *Young John Muir*, 29.
4. Holmes, *Young John Muir*, 39–72.
5. Quoted in Peter Wild, *Pioneer Conservationists of Western America* (Missoula, Mont.: Mountain Press, 1979), 32–34; Carolyn Merchant, *Reinventing Eden: The Fate of Nature in Western Culture* (New York: Routledge, 2004), 159–60.
6. Michael P. Cohen, *The Pathless Way: John Muir and American Wilderness* (Madison: University of Wisconsin Press, 1984), 5, 9–27.
7. Quoted in Wild, *Pioneer Conservationists*, 42.

8. H. Fitch to James Pinchot, 9 November 1849; James Pinchot to C. C. D. Pinchot, 16 August 1852, 1 October 1852, 5 November 1854, 13 March 1855, 16 March 1855, 18 April 1861, 30 June 1863; James Pinchot, General Letter, 12 July 1858, all in Gifford Pinchot Papers, Library of Congress (hereafter cited as GP).

9. Wallace to C. C. D. Pinchot, 15 April 1834, 6 May 1834, 5 May 1835; John Brodhead to C. C. D. Pinchot, 19 May 1837, 17 May 1837, 9 August 1839, 30 August 1839, 19 October 1845; Henry Fitch to James W. Pinchot, 9 November 1849; George Frost to James Pinchot, 17 May 1880; B. I. Stevens to James Pinchot, 9 December 1886, GP; Thomas R. Cox, "Transition in the Woods: Log Drivers, Raftsmen and the Emergence of Modern Lumbering in Pennsylvania," *Pennsylvania Magazine of Biography and History* 104 (July 1980): 345–64; Robert K. McGregor, "Changing Technologies and Forest Consumption in the Upper Delaware Valley, 1780–1880," *Journal of Forest History* 32 (April 1988): 69–81; Thomas Cox et al., *This Well-Wooded Land: Americans and Their Forests from Colonial Times to the Present* (Lincoln: University of Nebraska Press, 1985), 111–90.

10. James Pinchot to Mother, 21 August 1871; Mary Eno Pinchot, 1880 Memobook, GP.

11. James W. Pinchot, "The Yale Summer School of Forestry," *World's Work* (October 1904), 1–7.

12. Gifford Pinchot, *The Fight for Conservation* (New York: Doubleday, 1910), 38.

13. Pinchot, *Breaking New Ground*, 4th ed. (Washington, D.C.: Island, 1998), 103.

14. Pinchot, *Breaking New Ground*, 22. See also Pinchot, "Government Forestry Abroad," *Publications of the American Economic Association* 6 (May 1891): 191–238; Samuel P. Hays, *Conservation and the Gospel of Efficiency: The Progressive Conservation Movement, 1890–1920* (New York: Atheneum, 1969).

15. Clayton R. Koppes, "Efficiency, Equity, Esthetics: Shifting Themes in American Conservation," in *The Ends of the Earth: Perspectives on Modern Environmental History*, ed. Donald Worster (Cambridge: Cambridge University Press, 1988), 234–35; Char Miller, "Back to the Garden: The Redemptive Promise of Sustainable Forestry," *Forest History Today* (Spring 2000), 16–23.

16. Cohen, *The Pathless Way*, 320; William Frederic Badè, *The Life and Letters of John Muir* (Boston: Houghton, Mifflin, 1923), 2:265; Paul Russell Cutright, *Theodore Roosevelt: The Making of a Conservationist* (Urbana: University of Illinois Press, 1985), 249.

17. John Muir to Louise Muir, 13 June 1893, in Badè, *Life and Letters of John Muir*, 2:265.

18. Muir to Pinchot, 16 April 1894, GP; Pinchot to Muir, 19 June 1893, 17 September 1893, John Muir Papers, University of the Pacific (hereafter cited as MP).

19. Muir to Pinchot, 16 April 1894, GP; Pinchot to Muir, 13 September 1893, MP.

20. Pinchot to Muir, 13 September 1893, 8 April 1894, 23 May 1894, MP.

21. Muir to Pinchot, 16 April 1894, GP.

22. Cohen, *The Pathless Way*, 317–20; Michael L. Smith, *Pacific Visions: California Scientists and the Environment, 1850–1915* (New Haven, Conn.: Yale University Press, 1987), 159–61.

23. Smith, *Pacific Visions*, 160–61; Pinchot, *Breaking New Ground*, 101, 91.

24. Pinchot to Muir, 23 July 1896, MP.

25. Linnie Marsh Wolfe, ed., *John of the Mountains: The Unpublished Journals of John Muir* (Madison: University of Wisconsin Press, 1979), 357; Pinchot, *Breaking New Ground*, 100–103; Muir to Pinchot, 28 October 1896, 17 December 1897, GP.

26. Pinchot to Muir, 2 July 1897, MP; Muir to Pinchot, 8 July 1897, GP; Charles S. Sargent to Muir, 28 October 1897, MP.

27. Pinchot, *Breaking New Ground*, 119–22; Cohen, *The Pathless Way*, 292–94; Stephen R. Fox, *The American Conservation Movement* (Madison: University of Wisconsin Press, 1986), 112–15; Pinchot to Muir, 21 October 1896, MP.

28. Sargent to Muir, 3 May 1897, MP.

29. Muir, "A Plan to Save the Forests," *Century Magazine* 49 (February 1895): 630–31; Muir, "The American Forests," 336–42, and "The Wild Parks and Forest Reservations of the West," 1–36, both in *Our National Forests* (Boston: Houghton, Mifflin and Company, 1909); Cohen, *The Pathless Way*, 297–301.

30. Pinchot, *Breaking New Ground*, 133–39; Andrew Denny Rogers, *Bernhard Eduard Fernow: A Story of North American Forestry* (Princeton, N.J.: Princeton University Press, 1959); Stephen Ponder, "Gifford Pinchot: Press Agent of Forestry," *Journal of Forest History* 31 (January 1987): 26–35; Char Miller, "The Prussians Are Coming!" *Journal of Forestry* 89 (March 1991): 23–27, 41.

31. Koppes, "Efficiency, Equity, Esthetics," 233–38.

32. Koppes, "Efficiency, Equity, Esthetics," 233–38.

33. Pinchot, *Breaking New Ground*, 392–93.

34. Roderick Nash, *Wilderness and the American Mind* (New Haven, Conn.: Yale University Press, 1982), ch. 10; Muir to Robert U. Johnson, 23 March 1905, MP.

35. Pinchot to William E. Colby, 17 February 1905, GP; Muir to Pinchot, 27 May 1905, MP.

36. Pinchot to Marsden Manson, 28 May 1906, 15 November 1906, MP; *Independent* (8 August 1910), 375–76.

37. Muir to James Garfield, 6 September 1907, MP; Pinchot to Muir, 6 September 1907, MP; Pinchot to Roosevelt, 11 October 1907, quoted in Nash, *Wilderness and the American Mind*, 164.

38. Pinchot to Frederick Perry Noble, 18 September 1913, MP; Nash, *Wilderness and the American Mind*, 170–71.

39. Muir to Johnson, 11 September 1913, 3 September 1910, MP.

Nine

Gender and Wilderness Conservation

Kimberly A. Jarvis

On a cold September day in 1928, people gathered to dedicate the Franconia Notch Forest Reservation and Memorial Park to those who had served New Hampshire in times of war. In a cooperative effort by government and private groups, $400,000 had been raised from more than 15,000 contributors to purchase 6,000 forested acres on the western slopes of New Hampshire's White Mountains, thereby saving the region from purchase by logging companies. The New Hampshire Federation of Women's Clubs played a significant role in publicizing the Franconia Notch campaign and in its fundraising efforts. During the ceremony, Lula Morris, president of the New Hampshire Federation, noted that "it can be no secret that the New Hampshire Federation of Women's Clubs played a very real and important part in the final act that made it possible to acquire these thousands of acres. . . . the women of New Hampshire did not fail their state."[1]

Morris's comments highlighted one of the critical roles that women played in the American conservation movement during the last decades of the nineteenth century and the first decades of the twentieth century. Women dedicated their time and financial support to promote conservation campaigns that preserved places of unique scenic and historic beauty, protected wildlife, and advocated for the wise use and conservation of natural resources. While women's understanding and appreciation for their own connections to nature had its basis in part in their involvement with the outing and nature clubs of the late 1800s, their conservation ethic also developed out of their culturally mandated roles as nurturers and as protectors of the heritage and the future of the United States.

Until the latter decades of the twentieth century, most scholarly and popular discussions of the conservation movement focused on the men who led the movement at national, regional, and local levels. Historians of the conservation movement analyzed the contributions and leadership roles of Theodore Roosevelt, Gifford Pinchot, John Muir, Ralph Waldo Emerson, and Henry David Thoreau, among others, while leaving out key women, from Susan Cooper to Mary King Sherman to Rachel Carson. Roderick Nash's *Wilderness and the American Mind*, which in many ways defined conservation history for decades, was typical in this regard, with an almost complete absence of women. These studies of conservation organizations and campaigns led by men offered the impression that women did not participate in the conservation movement, and they were typical of much historical scholarship prior to the 1970s that ignored or devalued the contributions of women. By 1984, though, historian Carolyn Merchant began to identify many of these women conservationists and noted that it was women's efforts that extended the appeal of the primarily elite, male-led conservation movement and made it a mass movement. At the most basic level, recognizing women's contributions to the American conservation movement is essential in understanding the history of that movement.[2]

Even further, though, the politics of wilderness between 1870 and 1930 were in many ways linked to gender politics. Although the idea of wilderness was attractive to both men and women, wilderness was often referred to in masculine terms. Rugged and dangerous, wilderness experiences required physical strength and endurance, which would counterbalance the effects of the more effeminate modern urban life that many Americans believed threatened American masculinity. Contact with wilderness in the form of the American frontier had ensured the strength and development of the masculine American character. Saving wilderness, then, was also saving American manhood and, by extension, the nation.

In contrast, many Americans believed that women, seen as physically delicate, modest, and retiring, were not up to the challenges of wilderness. Women's place was in the private world of the home, away from the dangers and challenges of a vigorous and public life. Yet, paradoxically, in preservation campaigns throughout this period, women, if not the leaders of the movement (and sometimes they were) were almost always the bulk of the foot soldiers, donating time, money, and letters to the cause. Many women wilderness advocates used a typically Progressive Era politics based on women's perceived role as the caretakers of the nation to justify their activism. Insofar as America was "nature's nation," as seen earlier in this volume, protecting wilderness was protecting the nation and ensuring that people would continue to have access to the sacred and sublime characteristics of the wilds. Further, although the ideal of feminine fragility had no place at the top of the mountain, many women were members of outdoor clubs, went mountain climbing and camping, and participated in other wilderness activities. By recognizing that wilderness belonged to them too, women used the late nineteenth-century fascination with physical fitness and their role as the nation's nurturers and moral guides to justify their involvement in the conservation movement.

Women and Nature Conservation

American women's relationship with nature changed throughout the nineteenth century, reflecting the evolving conservation ethic as well as the larger societal changes that shaped the United States during this time. A significant illustration of these changing ideas is found in *Rural Hours*, published in 1850 by Susan Fenimore Cooper (daughter of the novelist James Fenimore Cooper). Cooper's book, a journal describing the changing of the seasons around her upstate New York home, reflected the literary efforts of other antebellum female nature writers who believed that nature study was a useful and beneficial way to spend their leisure. A well-read naturalist, Cooper's views of nature mirrored contemporary ideas and a traditionally feminine sensibility. *Rural Hours* helped to popularize the nature essay, and Cooper's work was known to the literary figures of the day, such as William Cullen Bryant, Washington Irving, and Henry David Thoreau (whose similar interest in studying nature was detailed by Bradley Dean in chapter 5). Cooper wrote of the interconnectedness of humans and nature through such seasonal events as the return of the robins and the excitement people feel about this sign of spring. Cooper's description of groves of old pines recognized the "wild, stern character of the aged forest pines" and the old trees' connections to the passage of time. Preserving unique aspects of nature, she argued, was the moral responsibility of those who valued the past.[3]

Cooper's *Rural Hours* suggested that there was a connection between nature and women's nurturing and protective characteristics. These instincts, as later defined by late nineteenth-century women, enabled them to become a force in the progressive conservation movement. Within the context of the conservation movement itself, the idea of conservation was used to demonstrate, and preserve, middle-class women's roles in society as nurturers and as the moral guides for the next generation. Many women who participated in conservation efforts wanted to ensure that their children, and the children of other women, would not only have resources for the future but also that there would be beautiful nature preserved for them to enjoy and as a source of their spiritual well-being.[4]

The conservation movement reflected the ideas behind a variety of reform programs and agendas that developed during the Progressive Era (c. 1890–1920). Many of these reforms attempted to control the increasing chaos of American society. By the 1880s and 1890s, it was becoming clear that the laissez-faire economic practices of the United States, responsible in part for its growth and prosperity, had also resulted in both social and environmental dislocations. Industrialization encouraged the rapid growth of urban areas and resulted in overcrowded living conditions and overtaxed urban infrastructures. By the turn of the twentieth century, urban reformers, primarily white and middle class, and many of them women, focused their attentions on the increasingly dismal conditions of factories and cities, "forc[ing] women to recognize the claims of society to their best thought and endeavor."[5] Through "municipal housekeeping," as such work came to be known, women gained insight into politics and found a respectable outlet which allowed them to focus their nurturing instincts on the larger problems of society outside the home.[6] Clubwomen became involved in issues such as prison and education reform and in urban reform efforts that improved cities' infrastructure and factories' working conditions (these reformers included most famously Alice Hamilton, the founder of occupational medicine, and Jane Addams of Hull House). Simultaneously, concern for the diminishing natural resources of the United States inspired a variety of public campaigns and government programs on the local and the national levels, led by both men and women (in chapter 8, Char Miller detailed the national-level context of these Progressive Era conservation reformers).

The women who worked within the conservation movement, then, were a subset of the much larger group of women who numerically dominated many of the late nineteenth-century reform movements associated with the Progressive Era, and they shared many of the same concerns and characteristics. These women had the time, the financial means, the motivation, and the necessary skills to effectively bring about social and political changes. The shrinking size of the American family along with the increasing affluence of the middle and upper classes during the latter years of the nineteenth century offered women as well as men the opportunity to participate in activities outside the home. Women employed their organizational abilities, developed during decades of charitable and reform work, to draw on grassroots support for the conservation movement. When women conservationists lobbied for the creation of parks and wilderness areas and for the preservation of the lives and habitats of various wildlife species, they could draw upon networks of similarly situated women across the country.

One outlet through which women were able to articulate their concerns for nature and for the future was the variety of women's clubs and organizations that developed between the 1860s and 1890s. The nineteenth-century women's club movement offered middle-class "ladies" local organizations though which they could improve themselves through study and social programs. Initially designed as self-help and self-education gatherings, by the late nineteenth century, these clubs began to focus not only on literature and the arts, but on local and national reforms as well, expanding their sense of community responsibility in an acceptable and uniquely feminine way. It was a small step from improving themselves to improving the larger community. While many women's clubs were local in nature, the founding of the General Federation of Women's Clubs in 1890 established a national organization that could look to state and local women's club networks to support a national agenda, including conservation causes. The General Federation indicated an interest in conservation in 1896 and created a Forestry Committee in 1902, but as early as the mid-1890s women's clubs in New York, Maine, and New Hampshire dedicated themselves to conservation efforts, including the creation of urban parks and green spaces.[7]

Women became leaders of conservation campaigns not only within their own club networks, but also within organizations that had male and female members. The Audubon Society and its campaign to save the snowy egret from extinction offer one example. Men and women who were interested in birds and birding in late nineteenth-century America could join the Audubon Society, founded by George Grinnell in 1886. After this first organization fell apart in 1888 under the weight of its own rapid expansion, Harriet Hemenway of Boston revived it in 1896. Gathering prominent members of Boston society about her, Hemenway and the Audubon Society utilized social networks to "work for reforms in the feather trade and for bird protection in general."[8]

The group organized a boycott of egret feathers, the long white plumes that were frequently part of the millinery designs of the late nineteenth century. The plumes were at their most beautiful during the egrets' breeding season, and hunters often killed the parent birds at their nests just after their young had hatched. Hemenway and other supporters saw the egret as a victim of human vanity. The boycott was successful; that the egret still exists today is due in part to the dedication of women like Hemenway and to the organizational power of groups like the Audubon Society.

Clubwomen worked together, utilizing established networks to reach their conservation goals, whether it was to preserve natural scenery or historic structures, or to protect the headwaters of major river systems in their state or on the national level. Through their efforts, these dedicated organizers and lobbyists for the conservation movement contributed to the creation of several state and national parks during the first decades of the twentieth century. Some women—Mrs. Lovell White, Alta McDuffee, and Mary King Sherman—were so much a part of the campaigns they organized that their names in some cases became synonymous with the land preservation they championed.

White's name came to define the efforts that resulted in the creation of Calaveras Grove State Park in California (1900–1954). The analysis of this campaign by historian Carolyn Merchant provided evidence as to women's skills in working within a political system still dominated by men.[9] White was the key organizer in the campaign to protect the Calaveras Grove of Big Trees, *Sequoia gigantean*, located on the western slopes of the Sierras approximately fifty miles north of Yosemite National Park. The grove was in danger of being logged by a timber company. When the San Joaquin Valley Commercial Association made inquiries as to how the region could come under either state or federal protection, White, the president of the California Federation of Women's Clubs, offered her organization's assistance.[10]

The California Federation of Women's Clubs passed a resolution that declared that "men whose souls are gang-saws are meditating the turning of our world-famous Sequoias into planks and fencing worth so many dollars." The destruction of the trees would be detrimental to the health of the people of California, as they were located in the Stanislaus River's watershed. It was "better [to have] a living tree in California than fifty acres of lumberyard. Preserve and plant [the trees] and the State will be blessed a thousandfold in the development of its natural resources." The club's appeals to state-level women's club federations throughout the country and to the General Federation of Women's Clubs resulted in "an avalanche of protests and petitions descending upon Senators and Congressmen from every state, calling their attention to the need for immediate action."[11]

In 1900, in response to the overwhelming public support for the preservation of the Big Trees, the owner of the land, R. B. Whiteside, left the grove "untouched" so that state or federal action could make the region into a park. In February of that year, White enlisted the aid of a California Federation of Women's Clubs vice president, A. D. Sharon, who was in Washington, D.C., when the campaign began. Sharon interviewed members of California's congressional delegation about the possibility of a Calaveras Grove park and even appealed in person to President William McKinley. A congressional bill in favor of acquiring the Big Trees was passed and signed by the president in March 1900. The gold pen that signed the bill was presented to the California Federation of Women's Clubs.[12]

The celebrations were short-lived, however. The congressional bill only authorized negotiations for the purchase of the grove; it had not allocated any funds for the purchase, and the bill to do so, presented to Congress in 1903, failed. White and her sister California clubwomen then organized a national campaign that netted 1.5 million signatures in support of legislation to create a national park around the sequoias. When the petition was sent to Washington, White noted that "this is the first instance on record where a special message has been sent to Congress at the request of an organization managed by women."[13]

In response to the petition, President Theodore Roosevelt sent a message to Congress that declared, "[T]he California Big Tree Groves are not only a California but a national inheritance and all that can be done by the government to insure its [*sic*] preservation should be done." It did little good. Not until 1909 did Congress authorize the exchange of national forest land for that of the grove, providing that both tracts "be of value substantially equal." No land was ever found. In 1926, the northern section of the Calaveras Grove was still under the ownership of Whiteside and still uncut, but the southern section of the grove was sold to another timber company. It would take another thirty years and the combined efforts of the newly formed Calaveras Grove Association (1926), the Calaveras Garden Club, and a "state-wide education campaign" before the Calaveras Big Trees State Park finally became a reality in 1954, over a half century after White's first efforts.[14]

White's commitment to conservation was not limited to her frustrating fight to preserve the Calaveras Grove. She was also president of the Sempervirens Club, founded in 1900. Begun over a campfire during a hike among the redwoods near San Jose, the Sempervirens Club's membership included men and women, college professors, sportsmen, clubwomen, nature lovers, and photographers, all of whom were dedicated to saving the other big trees of California, the coastal redwoods.[15] Another campaign, to save the Big Basin redwoods near Santa Clara, California, just south of San Francisco, suffered setbacks on the state level, but its success came much more quickly than that of the Calaveras Grove campaign. In 1901, the California state legislature authorized $250,000 for the purchase of 3,800 acres, creating Big Basin State Park, which was subsequently enlarged by an additional 5,200 acres.[16] White would go on to serve as forestry chair of the General Federation of Women's Clubs from 1910 to 1912. The Save the Redwoods League, which evolved from the Sempervirens Club in 1918, continues its work for the preservation of the coastal redwoods in the twenty-first century.

Although many of the most well-publicized early conservation campaigns were in the West, the women of the Northeast were also involved in conservation politics. The Maine Federation of Women's Clubs, the first state federation to join the General Federation in 1892, passed a variety of resolutions supporting the creation of a state forest reserve around Mount Katahdin, famously climbed and described by Thoreau (see chapter 5). The New Hampshire Federation of Women's Clubs participated in conservation campaigns and fundraising efforts that resulted in the creation of numerous state forest reserves and parks. Of these, the one that created Franconia Notch State Park in New Hampshire's White Mountains between 1923 and 1928 was the most successful.

In danger of being sold to timber companies after the resort hotel in the area burned down and was not rebuilt, Franconia Notch was an important source of tourist revenue as well as the home of the symbol of the state of New Hampshire, the granite profile of the Old Man of the Mountain. The campaign to protect Franconia Notch and the Old Man of the Mountain was a cooperative effort among the state of New Hampshire, the Society for the Protection of New Hampshire Forests, and the New Hampshire Federation of Women's Clubs. Led by Alta McDuffee, the New Hampshire Federation's campaign to save Franconia Notch drew on the support of state, regional, and national women's networks' support for conservation. Active in education and social welfare reforms in New Hampshire, McDuffee was a strong supporter of conservation issues during her 1921–1923 tenure as the New Hampshire Federation president and worked closely with Philip Ayres, the forester for the Society for the Protection of New Hampshire Forests, to ensure the success of the campaign.[17]

The campaign centered around efforts to raise $400,000 to enable the society to purchase 6,000 acres in Franconia Notch, which it would eventually turn over to the state. By the time McDuffee began organizing the New Hampshire Federation's campaign in January 1928, the state had contributed $200,000 and the Society for the Protection of New Hampshire Forests had raised $100,000. To make up the difference, McDuffee and the New Hampshire Federation asked people to contribute a dollar, in return for which they would symbolically "own" one of the trees located in Franconia Notch.

Promoted as a way to remember soldiers lost in World War I as well as a means through which to preserve an important part of American history, the Buy a Tree campaign brought in the remaining $100,000 through 15,000 contributions from across the nation. McDuffee publicized the campaign in the newsletter of the General Federation of Women's Clubs, telling the newsletter's editor that the New Hampshire Federation was "sure that women throughout the country are deeply interested in Franconia [Notch]."[18] McDuffee also traveled all over New England to discuss the campaign with gatherings of clubwomen. The Maine Federation of Women's Clubs endorsed the campaign, declaring that "no more worthy ambition could impress the minds of Club Women . . . than seeking to preserve inviolate, this valuable asset to posterity."[19] With the New Hampshire Federation's assistance, the remaining funds were raised in time to purchase Franconia Notch in June 1928. It was officially dedicated as a state park in September of that year.

McDuffee continued her conservation and club work through the next two decades, serving in the Society for the Protection of New Hampshire Forests as the assistant to the forester.[20] While in this position, McDuffee worked closely with several conservation projects and continued as editor of the New Hampshire Federation's newsletter. She later represented the New Hampshire Federation at the October 1947 ceremony when the society officially transferred the title of Franconia Notch to the state of New Hampshire.[21]

Women such as White and McDuffee provided leadership during important conservation campaigns aimed at preserving places of natural beauty. These women relied not only upon the efforts of local and state clubwomen, but also upon an organizational network of clubwomen at the national level. The General Federation of Women's Clubs spearheaded this network and its agenda. While many of the women's clubs focused on conservation or preservation issues in their own states, the General Federation's national agenda also influenced many of the resolutions passed by the state federations and many of the projects undertaken by federation members. In 1905, the *Federation Bulletin*, the General Federation of Women's Clubs' monthly newsletter, was pleased to note that "thirty seven State Federations have organized Forestry Committees . . . and the Chairmen [*sic*] are enthusiastically spreading the propaganda of tree-planting and forest perpetuation." The General Federation outlined a forestry and conservation agenda for its state federations that included working toward the appointment of a state forester in every state, the "introduction of some instruction of forestry into every school," and "the creation of State Forest Reserves."[22] The General Federation also voiced its support for the "legislation looking to the securing of Federal Reserves in the Southern Appalachian Mountains" in order to conserve timber resources and to prevent erosion at the headwaters of eastern river sources. The General Federation's goal through introducing guidelines for its forestry program was to "nationalize our interests and sympathies until the special work of each State becomes the general work of all States."[23]

Mary King Sherman, one of the General Federation's greatest proponents of these and other conservation efforts, utilized the same networks as other clubwomen in order to accomplish a more nationally oriented conservation agenda. She was an active member in the General Federation from 1904 until her death in 1934, serving in a variety of leadership positions, including recording secretary during the administration of Sarah Platt Decker, an avid conservationist, and, later, president of the General Federation. Sherman's interests combined a desire to preserve American wilderness with preserving the American home. As chair of the General Federation Committee on the Home through the early 1930s, Sherman and the General Federation lobbied successfully to have "homemaker" included as a category of employment on the U.S. Census. Conservation, however, was one of the most important issues during her first years with the General Federation, which earned her the sobriquet the "National Park Lady."[24]

Sherman's devotion to the preservation of nature dated to the three years that she spent with her son in her family's mountaintop cabin in Estes Park, Colorado. Bedridden by recurring malaria caught during a 1904 visit to the Panama Canal Zone while serving as a representative of the General Federation of Women's Clubs and by severe shoulder and back trauma, Sherman had little to think about, she said in a 1924 interview, besides the mountains and sky that were outside her window. In 1913, when she was finally able to climb to the top of Long Peak accompanied by her son, Sherman recalled:

> When after all I did reach the spot where the whole glory of the mountain ranges breaks upon you in all their magnificence—something wonderful did happen. It was all so much more glorious than I had dreamed that I was fairly overcome by emotion. I had lived almost all my life in flat country—and though I had loved natural beauty, I had no conception of such splendor as this. It was a profound revelation. And as I sat there, drinking it in, I made a solemn vow that if I were able to make the return trip in safety, I would devote the rest of my life to helping others see and feel what I had experienced—the vast beauty of the world. I pledged myself to help save such scenes as this for that purpose.[25]

Sherman's "revelation" was a near-religious experience that could have as easily been described by any romantic worshiper of the wilds described earlier in this volume—and she immediately moved from spiritual insight to a political commitment to preserve this unique natural beauty so that others could be transformed by the wilderness.

Inspired by her mountaintop experience, Sherman returned to her work with the General Federation of Women's Clubs. She was appointed chair of the Conservation Department in 1914. She reorganized the department, revamping old programs and developing new literature. She was instrumental in lobbying Congress for the creation of Rocky Mountain National Park in Colorado in 1915. She represented the General Federation of Women's Clubs at the park's dedication ceremony.

Through Sherman's efforts, the support of the General Federation of Women's Clubs was thrown behind the creation of the National Park Service in 1916. She supported the successful 1916 campaign for Grand Canyon National Park, which had been in the works for thirty-three years. By 1920, she had been involved in lobbying for six national parks, as well as fighting for the independence of the National Park Service from the "great water and irrigation interests" that threatened the movement for park conservation. Her efforts were recognized by an honorary lifetime trusteeship in the National Park Service.

Sherman brought women's work in conservation onto the national stage. She recognized, as did the General Federation of Women's Clubs, the potential power in organizing women throughout the United States. State federations could call upon their sister organizations when support was needed, and national campaigns could be more effectively carried out on the state and local levels if the state federations were both aware of and in accord with the national group. Sherman's dedication to the conservation movement and to the creation of parks that preserved unique scenic beauty was echoed throughout the United States. The success of these conservation campaigns drew upon the impulses of clubwomen, whose sense of social responsibility for the welfare of the nation and for that of future generations contributed to the unity of purpose shared by clubwomen across the nation. And in their conservation politics, they drew upon social gender norms of women as "municipal [national] housekeepers" and argued that saving wilderness was keeping the nation's house in order.

Men and Women in the Mountains

There were many people, both men and women, who, like Sherman, wanted to experience the "profound revelation" of nature for themselves. John Muir spoke for many when he wrote that "mountain parks and reservations are useful not only as foundations of timber and irrigating rivers, but as fountains of life."[26] Sherman, Muir, and their supporters felt that there needed to be places of quiet natural beauty, where people could escape the anxiety and dirt of city life. These ideas reflected, in part, a larger cultural fascination with wilderness (and what it embodied) that was common to many Americans during the last decades of the nineteenth century. This "wilderness cult" represented nature as a retreat from the overcivilized cities that offered the opportunity to experience untamed landscapes and the chance to see the grand, unique scenery that defined American exceptionalism.[27] The return to wilderness also offered an opportunity to recapture the essence of the American character that had developed in response to the most elemental wilderness in American history: the frontier.

Middle- and upper-class reformers and intellectuals gained a new appreciation for wilderness as it became associated with ideas about the unique qualities of American history and character. Wilderness and nature were viewed as settings where Americans could find traces of what had once and continued to define them—the ruggedness of a frontier existence. It was in this contact with the savage and untamed wilderness that the American character found its definition. In 1893, Frederick Jackson Turner declared in "The Significance of the Frontier in American History" that it was through the idea of the frontier that "American development" could be explained. Turner argued that "to study this advance, the men who grew up under these conditions, and the political, economic and social results of it, is to study the really American part of our history," thereby establishing the idea of the frontier as central to American history. Contact with the frontier kept "alive the power of resistance to aggression, and developing the stalwart and rugged qualities of the frontiersman," each distinctly masculine characteristics. (In fact, Kit Carson's mother is the only woman mentioned in Turner's essay.)[28]

By the turn of the twentieth century, the self-discipline and self-reliance that characterized those earlier generations of pioneering American men seemed to be disappearing in the face of the corruption and the overcivilization that characterized middle-class urban life. The societal and psychological conflicts associated with a rapidly expanding industrial society were new and troubling as well.[29] In addition, the influx of approximately twenty million immigrants between 1880 and 1920 increased white native-born Americans' anxieties about the dilution of the American character through inferior stock and foreign traditions and influences. Similar fears had motivated many middle-class women reformers to focus upon one potential response: cleaning up and "Americanizing" immigrants and their homes, as in the settlement house movement. Another possible antidote to these concerns was the idea of "the strenuous life."

The origins of the phrase, which is associated with the vigorous, active existence promoted by Theodore Roosevelt and his contemporaries, lay within ideas about manliness, civilization, nationalism, imperialism, and racism. Roosevelt first used the term in an 1899 speech to encourage the United States to pursue imperial goals and therefore encourage the progress of the dominant white, Anglo-Saxon race. In this way, the strenuous life, which would reinvigorate American society in part through its connection to wilderness, could further the progress of the United States through expansion overseas.[30]

For privileged urban men like Roosevelt, the introduction to the frontier experience often began in childhood through popular nineteenth-century novels about adventures with wild animals or Indians as well as through contact with wilderness itself through travel. Both experiences encouraged the development of a civilized man, who had to be brave and self-reliant, patient and enterprising. Roosevelt's goal was to cultivate the masculine, rugged, and independent qualities of the frontiersmen in the civilized gentlemen who lived and worked in America's cities. Founded by Roosevelt in 1887, the exclusive, all-male Boone and Crockett Club, which worked to prevent the slaughter of game animals by commercial hunters (yet required that prospective members have killed several trophy animals themselves) also promoted the ideals of manliness and a robust American nation for elite urban men. Obviously, the backwoods hunters described in Ben Johnson's chapter in this volume would not have been invited to join.

Although the strenuous life was thought of as a masculine virtue, the late nineteenth century also witnessed a more general growth in interest in physical fitness. Women, as well as men, were encouraged to be more active, and even women's everyday clothing changed, becoming less confining and allowing for more freedom of movement. The emphasis on physical fitness and outdoor life was evident, too, in the creation of the Boy Scouts (1910) and the Girl Guides (1912, later the Girl Scouts). While the Boy Scouts offered a means through which American boys could retain their connections to the "outdoor life" and through experiencing the wilderness learn to be men, the Girl Scouts also included a significant focus upon nature lore and fitness, as well as on domestic skills.[31] The strenuous life was not just being enjoyed by men.

For adults, outdoor groups—particularly hiking clubs—advocated the benefits of being in and preserving nature, and membership in many of these organizations was open to both men and women. Through groups such as these, men and women (usually elite urban professionals) had contact with wilderness, which helped to raise their awareness of its beauty and significance, even as they participated in the physical wilderness challenges of the imagined frontier. Organizations such as the Appalachian Mountain Club (founded 1876) were inspired by some of the same ideas that influenced the conservation movement and gave active men and women the opportunity to encounter nature firsthand with other like-minded people. Through its trail system in the eastern United States, the Appalachian Mountain Club offered access to wilderness and the chance to experience the primitive in an area that was becoming increasingly crowded and commercialized. Members of the Appalachian Mountain Club also brought awareness to the Northeast that there were local areas of scenic beauty or historic value that should be set aside and preserved, and private groups, as well as local and state governments, could work to save these areas. By 1890, the group had added to its charter the mission of preserving places of special scenic or historic importance.[32]

The Appalachian Mountain Club's early and continued success was due in part to its decision to include women in its membership. Women's participation with hiking in the eastern mountains dated back at least to 1838, when Benjamin Silliman, editor of the *American Journal of Science and Arts,* noted that "ladies sometimes go on this adventure," referring particularly to the ascent to the 6,822-foot summit of Mount Washington in the White Mountains of New Hampshire. In Silliman's "judgment," however, women should not attempt it "because of the fatigue that resulted from the arduous climb" and the fact that they must make the trip "unaided." If women did "insist on making this ascent, their dress *should be adapted to the service* and none should attempt but those of firm health and sound lungs." Silliman did make the point, however, that all climbers, both men and women, should be healthy and able to handle the climb.[33]

Although Silliman was doubtful of women's ability to climb mountains, the Appalachian Mountain Club did not share his concerns. In its second meeting, the club voted to allow women to join, and from then on the club's membership included an increasing number of women interested in hiking and exploring. Between 1876 and 1886, roughly 10 percent of the club's membership was women. In 1876, there were 119 members, of whom 12 were women. By 1886, 67 out of 715 members were women.[34] These women almost immediately began contributing articles to *Appalachia*, the club's newsletter. In 1877, W. G. Nowell wrote about "A Mountain Suit for Women," emphasizing the need for a "simpler costume," as women's dress "has done all the mischief. For years it has kept us away from the glory of the woods and the grandeur of the heights. It is time we should reform." Nowell used her own experience to describe the cumbersome nature and danger of traditional skirts: her skirt caught on the corner of a rock and she was almost thrown into a ravine. The suit, Nowell decided, should be feminine but needed to be practical and safe as well.[35] Later articles by other women described sensible advice for short trips and camp life, and women continued to contribute articles about hiking in a variety of locations. The women of the Appalachian Mountain Club participated in the same activities as the men, although none of them appears to have served as an officer within the first two decades of the group's existence.

The Sierra Club, founded in California in 1892, followed the same basic principles as the Appalachian Mountain Club. Its original membership, which included John Muir, was similar to that of the Appalachian Mountain Club. Both hiking and wilderness preservation were part of the club's mission. Women were involved in the Sierra Club's activities from its first outing, which included a group of female college students from Berkeley and Stanford.[36] Women continued to avidly participate in club outings and activities, and between 1927 and 1929, Aurelia Harwood served as the Sierra Club's first woman president.[37]

Women's involvement with outdoor clubs offered them the opportunity to experience wilderness in the same way that men did through hiking, camping, and direct contact with nature. Women's accounts of their wilderness experiences echoed those of Mary King Sherman when she reached the top of Estes Peak after her long illness. Their writings indicate that they experienced a sense of accomplishment, wonder, and a spiritual connection to nature that mirrored the reactions of many of their male colleagues.[38] Not surprisingly, women's increasing participation in mountain climbing and hiking led to a change in the way that some people understood gender and, particularly, the relationship between wilderness activity and masculinity. Opponents of wilderness activists drew attention to these changing gender norms among wilderness supporters, as when one 1910 critic called the opponents of the Hetch Hetchy dam "short-haired women and long-haired men."[39] This epithet still resonates today in some critiques of contemporary environmentalism. Throughout the twentieth century, some opponents of wilderness preservation have harbored the suspicion that men who seek to preserve natural beauty instead of promoting national development are effeminate, while women who climb mountains are masculine, and that both are subverting American national strength. This highlights the continued relevance of gender in understanding wilderness politics in an America where there is still a cultural resonance to the notions of masculine struggle in the frontier and feminine preservation of beauty, where more women than men join environmental organizations, and where more men than women participate in "extreme" outdoor sports.

Ideas about wilderness as a frontier proving ground, as a place of spiritual renewal, and as a source of prosperity for the present and future intersected with ideas about the roles of men and women in American between 1870 and 1930. Men and women often saw wilderness in similar ways—as the means to define American exceptionalism, as a place of challenge and recreation, as a refuge from the chaos of industrial America. Those who shaped the conservation movement's initial focus drew upon a variety of influences, including Theodore Roosevelt's promotion of the reinvigoration of American society through a decidedly masculine, aggressive nationalism and imperialistic policy and through renewed contact with nature and, more specifically, wilderness. The challenges of wilderness would keep America strong and vibrant, a place of rugged individualists. At the same time, clubwomen and other reformers worked to improve American society and to conserve its resources so that future generations would continue to benefit from its prosperity. The conservation movement, which initially reconciled the later divisive ideas of conservation of resources versus preservation of nature, drew on both of these influences, combining the idea of rugged wilderness with a sense of moral responsibility to future generations.

The relatively recent investigation into women's place in conservation history has added a richness to the discussion of gender's influence on the movement itself and to the larger discussion of the image of wilderness in late nineteenth- and early twentieth-century America. The white middle-class and elite women who came to the conservation movement through nature study, women's clubs, outdoor organizations, or as reformers utilized the era's definitions of feminine character—specifically, women as moral nurturers—as the means through which to make impressive and decisive contributions to conservation campaigns. Their strength came through organization and through commitment to their ideals. Other women challenged those notions of the feminine character and took to the trails, similarly dedicated to their vision of the good and meaningful life. In so doing, these women of the American conservation movement made a place for themselves in the American wilderness.

Notes

1. Society for the Protection of New Hampshire Forests, "Addresses at the Dedication of Franconia Notch," September 15, 1928, Box 10, folder 6, Society for the Protection of New Hampshire Forests Collection, Milne Special Collections, University of New Hampshire Library, Durham, New Hampshire. Hereafter abbreviated as SPNHF Records.

2. Roderick Nash, *Wilderness and the American Mind*, 3d ed. (New Haven, Conn.: Yale University Press, 1982); Carolyn Merchant, "Women of the Progressive Era Conservation Movement, 1900–1916," *Environmental Review* 8 (1984): 57.

3. Vera Norwood, *Made from This Earth: American Women and Nature* (Chapel Hill: University of North Carolina Press, 1993), 25, 27, 41; Susan Fenimore Cooper, *Rural Hours: New and Revised Edition* (Boston and New York: Houghton, Mifflin, 1887), 194.

4. Norwood, *Made from This Earth*, 40; Merchant, "Women of the Progressive Era," 57, 73–75.

5. Mary I. Wood, *The History of the General Federation of Women's Clubs* (New York: General Federation of Women's Clubs, 1912), 30.

6. Suellen M. Hoy, "'Municipal Housekeeping': The Role of Women in Improving Urban Sanitation Practices, 1880–1917," in *Pollution and Reform in American Cities, 1870–1930*, ed. Martin V. Melosi (Austin: University of Texas Press, 1980), 173–98.

7. Karen Blair, *The Clubwoman as Feminist: True Womanhood Redefined 1868–1914* (New York: Holmes & Meier, 1980), 5; Priscilla G. Massmann, "A Neglected Partnership: The General Federation of Women's Clubs and the Conservation Movement, 1890–1920" (Ph.D. diss., University of Connecticut, 1997), 70–71.

8. Jennifer Price, *Flight Maps: Adventures with Nature in Modern America* (New York: Basic, 1999), 2–3, 63.

9. See Merchant, "Women of the Progressive Era"; and Carolyn Merchant, *Earthcare: Women and the Environment* (New York: Routledge, 1996), 109–36.

10. Mary S. Gibson, *A Record of Twenty Five Years of the California Federation of Women's Clubs* (n.p.: California Federation of Women's Clubs, 1927), 1:174.

11. Merchant, "Women of the Progressive Era," 59.

12. Merchant, "Women of the Progressive Era," 59.

13. Gibson, *A Record of Twenty Five Years*, 176.

14. Gibson, *A Record of Twenty Five Years*, 176.

15. Gibson, *A Record of Twenty Five Years*, 14–15; Merchant, "Women of the Progressive Era," 60.

16. Gibson, *A Record of Twenty Five Years*, 18.

17. Henry H. Metcalf, "Representative Women in New Hampshire: Mrs. Charles H. McDuffee," *Granite Monthly* 60 (March 1928): 226.

18. Alta McDuffee to Vella Winter, Washington, D.C., November 19, 1927, Box 9, folder 16, SPNHF Records.

19. Josephine Skofield, Portland, Maine, to McDuffee, December 28, 1927, Box 9, folder 16, SPNHF Records.

20. Ayres to Viola S. Smith, Boston, June 6, 1930, Box 11, folder 16, SPNHF Records.

21. Society for the Protection of New Hampshire Forests, "Annual Forestry Conference Program, Friday, November 3, 1947," Box 25, folder 47, SPNHF Records.

22. "Forestry," *Federation Bulletin* 3, no. 1 (October 1905): 13.

23. "Forestry," *Federation Bulletin* 3, no. 1 (October 1905): 13.

24. Frances Drewry McMullen, "The National Park Lady," *Woman Citizen* (May 17, 1924), 10; General Federation of Women's Clubs Archives, Presidents' Papers (Record Group 2), Papers of Mary Sherman.

25. McMullen, "The National Park Lady," 10.

26. John Muir, "The Wild Parks and Forest Reservations of the West," in *John Muir: Nature Writings* (New York: Literary Classics of America, 1997), 721.

27. Nash, *Wilderness and the American Mind,* 141–60.

28. Frederick Jackson Turner, *The Frontier in American History* (New York: Holt, Rinehart and Winston, 1920), 1, 4, 15.

29. T. J. Jackson Lears, *No Place of Grace: Anti-modernism and the Transformation of American Culture, 1880–1920* (Chicago: University of Chicago Press, 1984), 26.

30. Gail Bederman, *Manliness and Civilization: A Cultural History of Gender and Race in the United States, 1880–1917* (Chicago: University of Chicago Press, 1996), 184.

31. Nash, *Wilderness and the American Mind,* 147–48.

32. "Report of the Councillors," *Appalachia* 6 (July 1891): 260.

33. Benjamin Silliman, "Remarks by the Editor," *American Journal of Science and Arts* 34 (April 1838): 77.

34. Membership numbers were compiled from the numerous reports on membership anonymously published in their newsletter: "Membership Report," *Appalachia* 1 (1876–1878): 5–6, 126, 200, 300; *Appalachia 2* (1879–1881): 84–87,189–190, 295-296; *Appalachia 3* (1882–1884): 89–97, 200–202, 296–298, 389–390; *Appalachia 4* (1884–1886): 95, 175–176, 283–284, 383–385.

35. W. G. Nowell, "A Mountain Suit for Women," *Appalachia* 1 (June 1877): 181–83.

36. Stephen Fox, *The American Conservation Movement: John Muir and His Legacy* (New York: Little, Brown, 1981), 341–42.

37. Fox, *The American Conservation Movement,* 343.

38. Susan R. Schrepfer, *Nature's Altars: Mountains, Gender, and American Environmentalism* (Lawrence: University Press of Kansas, 2005), 68.

39. Nash, *Wilderness and the American Mind,* 169.

Ten

Putting Wilderness in Context
The Interwar Origins of the Modern Wilderness Idea

Paul Sutter

The American wilderness idea, as previous chapters have suggested, has a deep, complex, and contested history. But what I call the *modern wilderness idea*—the notion that Americans ought to preserve "wilderness areas" as a distinct federal land designation, a policy made law with the passage of the Wilderness Act of 1964—had its origins during the interwar years.[1] There were important threads from the national parks activism of the previous half century woven into interwar advocacy, but interwar activism was formative primarily because it led to the eventual statutory creation of a national wilderness system and because it produced many of the wilderness movement's canonical figures, including Aldo Leopold and Bob Marshall. Yet to see the interwar era merely as a prelude to postwar passage of the Wilderness Act is to miss the fascinating context that drove the creation of the modern wilderness system. The central argument of this chapter, then, is that we need to appreciate not only that interwar activism was the foundation for postwar wilderness politics, but that interwar events and trends played a formative role in shaping the modern wilderness idea. We need to understand interwar wilderness thinking on its own terms.

The modern wilderness idea emerged out of conditions different from those that had dominated Progressive Era wilderness politics. As Char Miller pointed out in chapter 8, the battle over the Hetch Hetchy Valley in Yosemite National Park highlighted, perhaps too rigidly, the opposing doctrines of utilitarian conservation and aesthetic preservation that guided the management of the national forests and national parks, respectively. While the preservationist cause, championed by John Muir, lost the day in Hetch Hetchy, it built strength in the larger battle over the fate of America's remaining public lands. Indeed, it is tempting to see interwar wilderness advocacy as merely an extension of the tide of preservationist sentiment that rose after Hetch Hetchy's loss. But the ascendance of preservationism was not only the result of the intellectual forcefulness of its proponents; changes in outdoor recreation, built around the automobile and good roads, meant that more Americans were interested in preserving—and developing—the public lands for their leisure. Among other things, those changes empowered the national parks lobby, resulting in the creation of a National Park Service in 1916 and effective campaigns for new national parks. But they also set the stage for the modern wilderness idea, not merely as an evolutionary step beyond the national park idea, but as a critical response to the ways in which preservation and recreational development increasingly went hand in hand. As large numbers of Americans set out in their automobiles in search of a wilderness experience, and as they demanded roads and tourist facilities consistent with their modern mode of transportation, another much smaller group of Americans—the wilderness advocates who are the subject of this chapter—began to argue that automobiles and a modern tourist infrastructure were antithetical to wilderness preservation.

There is something jarring about this argument. The traditional story has emphasized how wilderness proponents through the late nineteenth and early twentieth centuries crafted their advocacy in opposition to the economic forces—ranching, mining, timber cutting, agriculture—that were transforming the American landscape. By standing between these forces of production and the nation's wildlands, the story goes, these advocates saved crucial wilderness remnants to be enjoyed as recreational and leisure spaces. Moreover, the Hetch Hetchy conflict revealed that the ascendant utilitarian conservation tradition, which emphasized the wise use of natural resources, could also be a threat to wilderness. To the extent that the interwar era had a distinct character within this older narrative, it was as a period in which the science of ecology pushed preservationists beyond scenery to embrace a purer wilderness ideal. In the traditional telling, then, the interwar wilderness idea emerged as a more sophisticated tool for opposing not only transformative land and resource use, but also the anthropocentrism of utilitarian conservation and the aesthetic limitations of park preservation.[2]

While this interpretation has its merits, I argue that interwar wilderness advocates crafted their preservationist policy less to shore up opposition to resource use and the philosophical underpinnings of utilitarian conservation than to defend wildlands against the threats of motorists, road builders, the nascent forces of industrial tourism, and the government agencies outfitting the public lands for motorized outdoor recreation. Interwar wilderness activists formulated the wilderness idea largely to oppose modern *recreational* trends, not to offer recreational preservation as an alternative to resource development on the public lands. The modern wilderness idea was less a higher form of the national park ideal than it was a response to the compromises and tensions that were making park preservation politically attractive at the dawn of the automobile age.

When we place wilderness squarely within the context of the interwar period, we face the irony that one set of inheritors of the deeper American wilderness tradition documented thus far, recreational motorists, came to appreciate wilderness in such a way that it drove another set of inheritors into a more radical definition of wilderness. In the service of examining that irony, I briefly explore the terrain of interwar outdoor recreation before turning to four of the thinkers and activists—Aldo Leopold, Robert Sterling Yard, Benton MacKaye, and Bob Marshall—who embraced wilderness as a new preservationist model during the 1920s and 1930s and who came together to form the Wilderness Society, the first national organization dedicated to the preservation of wilderness areas. These four activists did not constitute the entirety of the interwar wilderness movement; there were other national figures, such as Rosalie Edge of the Emergency Conservation Commission, and a myriad of regional and local activists who fought for the preservation of America's wildlands and wildlife. Nor did they fully embody or successfully monopolize wilderness thought. The millions of recreational motorists who sought out the nation's wildlands during this period were themselves responding to, and in turn shaping, the wilderness idea as they understood it. But the founders of the Wilderness Society deserve our particular attention for two reasons: because they critically engaged with the ways in which roads and automobiles were transforming wild nature and American recreational habits and because their activism launched the campaign for passage of the Wilderness Act. Indeed, the Wilderness Society was the key organization that carried the modern wilderness idea into the postwar years. By focusing on these four figures, then, I narrow the focus of this volume to take a detailed look at the era's most distinctive and influential contribution to America's larger wilderness history.

The Automobile and Interwar Trends in Outdoor Recreation

New patterns of mobility reshaped American wilderness thought and politics in the early twentieth century. Automobile ownership grew slowly during the first decade of the century, but with assembly line production commencing in the mid-1910s, automobiles quickly became affordable to middle-class Americans. In 1910, there had been only 1 automobile for every 265 Americans; by 1929, the ratio was about 1 in 5.[3] The drivers of all of these cars needed good roads to traverse. In 1916, the same year that it created the National Park Service, Congress passed the first of a series of Federal Aid Highway acts, committing the federal government to funding road improvement and coordinating a national system of roads. Americans eagerly took to the nation's roads, improved and unimproved, in search of recreational nature. A small number attempted highly publicized cross-country trips in the first decades of the twentieth century, but a larger group of Americans pursued more modest forms of auto camping and motor touring, taking overnight and weekend trips into and through the American countryside. From the start, the automobile was a technology for getting back to nature.

Before World War I, auto camping was a chaotic affair, its practitioners plying hinterland roads and stopping wherever they pleased. Early auto campers often camped on private land and used—or abused—privately owned resources, leading many rural landowners to post their land against trespass. One result of these abuses was the growth of municipal and private auto camps and, later, motels, which provided for motorists' needs on an increasingly commercial basis. Another result was that those auto campers who wanted a wild, isolated experience moved their activities from local landscapes to the public lands, not only because automobiles and improved roads allowed them to get there but also because the private countryside was no longer a recreational option. The closing off of the rural landscape as a recreational space was an important force in propelling interwar American motorists greater distances in search of wild nature.[4]

During the 1910s and 1920s, Americans headed "back to nature" in large numbers not only because they had the technological capacity to do so, but also because cultural production—newspapers, magazines, postcards, advertisements, and promotional literature—encouraged nature tourism for a mass audience on a far broader scale than it had in previous decades. Entrepreneurs, civic boosters, and even the National Park Service embraced the campaign See America First during the 1910s and 1920s as a way to encourage Americans to see their own country instead of going on a European Grand Tour.[5] Beginning in the mid-1910s, park officials embarked on an ambitious publicity campaign to encourage Americans to get to know their parks as a patriotic duty, and booster groups in and around the parks also advertised their local destinations so as to profit from the tourist trade. After World War I, nature tourism and outdoor recreation became cultural imperatives, crucibles of national character that filled the vacuum created by the vanishing frontier.

Stephen Mather and Horace Albright, the most important early leaders of the park service, saw motor tourism as a source of an enlarged political constituency for park protection. Institution builders that they were, they lured motorists to the parks by building and improving roads and other tourist facilities. The automobile, they hoped, would democratize the park experience by providing more Americans with affordable access to the national parks, which to that point had largely been the domain of the wealthy. And, indeed, park visitation soared, from about 240,000 visitors in 1914 to more than 3.5 million by 1931.[6] Motor tourism, moreover, lent park preservation an economic rationale that had been lacking as late as the Hetch Hetchy conflict, when proponents of water development had maligned preservationists as selfish aesthetes who cared more about nature than economic progress. But amid all of this enthusiasm for the automobile and its benefits for national parks, some began worrying that the park service was too eager to develop its holdings for mechanized visitors and that such developments might come at the expense of a wilderness experience. The modern wilderness idea was a product of such concern.

The national forests—which had a much larger land base than the national parks—also saw an influx of motorized visitors from the 1910s on. But foresters were more ambivalent about this development than were their administrative counterparts in the park service. Auto campers came of their own volition, making use of an administrative infrastructure of roads and trails that foresters had built over the previous decades. Moreover, motorized recreational use of the national forests took hold at a time when there was little demand for public timber. Most commercial timber still came from private lands, and when federal foresters attempted to sell national forest timber, industry accused them of flooding the market and driving down prices. Unable to practice the sustainable forestry that was the essence of their training, federal foresters bided their time by building roads and trails, opening up remote national forests to their scrutiny and control in ways that inadvertently provided prime opportunities for recreational motorists. Moreover, a disproportionate amount of early federal road-building money went to national forest roads, not only to access timber resources and to protect them against fire, but also to connect western communities isolated by these federal holdings. As a result, new roads whittled away at the national forests' substantial roadless acreage more extensively than they did within the national parks, where, until the New Deal, the National Park Service modernized an existing infrastructure that predated the automobile.

The interwar years were marked by a bureaucratic rivalry between the National Forest Service and the National Park Service, with park service officials scheming to cherry-pick national forest properties for new national parks and forest service officials intruding upon the recreational territory of the park service. But forest service recreational planning and development was not entirely, or even primarily, motivated by such a rivalry. Recreational users dragged forest service officials reluctantly into such planning. In the years immediately after World War I, the forest service flirted with making recreation central to its mission, but for a number of reasons it balked at doing so. Embracing recreation would have confused a public still figuring out the differences between national forests and national parks, and it would have introduced a mission potentially at odds with growing trees. Although forest service officials defended their territory against park service usurpation, sometimes suggesting that they could manage key recreational sites better than the park service, it is more accurate to see early recreational management of the national forests as an initiative foisted upon foresters by a wave of motorized visitors than as the product of a rivalry.

Ironically, the development of a forest service wilderness policy in the early 1920s was more the result of that ambivalence about recreation than it was the result of growing recreational enthusiasm within the agency. Foresters on the ground scrambled, with few resources, to contain growing recreational use and the problems that came with it, such as sanitation crises and increased fire risks. They also responded to the public hunger for recreation with several indirect approaches that accommodated recreational use without prioritizing it. In some cases, they worked with local civic groups, which developed recreational areas on nearby national forests with their own funding and labor. More important, the forest service granted permits, through the Term Permit Act of 1915, which allowed individuals or groups to lease national forest lands and develop private recreational facilities on them—from summer camps and cottages to resort hotels. The result was a proliferation of private facilities in scenic areas of the national forests. Lacking the funding and the will to build public recreational facilities, the forest service opted for permitting private development of a sort that highlighted the need for wilderness preservation.

By the mid-1920s, with both the park service and the forest service contending with skyrocketing visitation and new patterns of mechanized use, President Calvin Coolidge convened the National Conference on Outdoor Recreation (NCOR), whose officials and delegates he charged with crafting a national outdoor recreation policy. The NCOR stood in marked contrast to the well-known 1908 Governors' Conference on Conservation, during which the utilitarian ideology dominated and hardly a word was uttered about recreation. The two NCOR meetings, in 1924 and 1926, were devoted to defining the public interest in outdoor recreation, and they were fractious and cacophonous affairs. Groups such as the American Legion boasted about the martial advantages of a nation whose citizens exercised in nature and developed outdoor skills (the military mobilization for World War I, in its camp training and supply of surplus materials made available after the war, did much to spur an interest in the outdoors); labor and urban groups argued for more and closer recreational outlets for workers and other city dwellers, whose hours on the job were slowly declining; the American Automobile Association and the U.S. Chamber of Commerce highlighted the commercial advantages of recreational tourism; the National Highways Association argued for the recreational benefits of road modernization; and sportsmen promoted hunting and fishing on federal lands, while defenders of wildlife and other preservationists hoped that a federal recreational policy would emphasize stricter preservation of federal lands and wildlife. Only at the second meeting would a voice for wilderness designation—Aldo Leopold's—be heard. If there were any clear lessons to be drawn from these meetings, they were, first, that outdoor recreation had become a widespread phenomenon invested with considerable cultural meaning, and second, that careful planning was necessary if the federal government hoped to protect opportunities for all of these various and sometimes competing recreational constituencies. Wilderness preservation emerged in this context as one of many such claims on public lands recreation.[7]

The Depression and New Deal represented a dramatic departure from the economic conditions that had shaped the outdoor recreation boom of the late 1910s and 1920s, but the boom itself continued with surprising strength, facilitated by federal programs dedicated to building recreational infrastructures and by an ideology of national recovery in which therapeutic nature played a starring role. New Deal programs such as the Civilian Conservation Corps (CCC) and the Public Works Administration (PWA) pumped millions of dollars into the coffers of the park service and forest service and provided them with tens of thousands of emergency conservation workers to build roads, trails, and campgrounds and to create entirely new federal and state park facilities. The New Deal was thus a climax to the nation's growing interest in recreational nature during the interwar years, and the public conservation work it accomplished set the stage for an even greater expansion of recreational interest and development in the postwar years. But interwar wilderness activists reacted with concern to this dramatic mobilization of federal money and labor. Indeed, in the first issue of their magazine, the *Living Wilderness,* the founders of the Wilderness Society bemoaned not only the threats that road building posed to remaining federal wilderness but also the New Deal impulse "to barber and manicure wild America as smartly as the modern girl."[8] To these advocates, the uncritical application of New Deal labor threatened the wildness of the public lands.

New Deal threats precipitated an organizational response to concerns about wilderness almost two decades old. The Wilderness Society had its informal founding by the side of a Tennessee road in the autumn of 1934, as delegates—among them Marshall and MacKaye—from the American Forestry Association's annual meeting traveled from Knoxville, seat of the newly created Tennessee Valley Authority (TVA), to inspect a CCC camp near Norris Dam. In January 1935, the Wilderness Society had its formal founding in Washington, D.C. Among its founding members, Aldo Leopold, Bob Marshall, Benton MacKaye, and

Robert Sterling Yard were the most important shapers of the modern wilderness idea that the Wilderness Society would be so crucial to promoting during the postwar years. In grappling with the interwar context, these four figures came, over the course of two decades, to embrace—and to redefine—wilderness as a modern policy idea.

Aldo Leopold

It was the recreational conditions he encountered in the national forests of the West that first led Aldo Leopold to propose wilderness preservation.[9] Born and raised in Iowa, Leopold earned his forestry degree from Yale in 1909 and then went to work for the U.S. Forest Service. The 1910s found him ranging over large stretches of rugged and roadless national forest territory in the Southwest. Leopold's passion was game policy, but he had to contend with other pressing aspects of national forest recreation. In 1916, he coauthored two telling reports: one on the rapid recreational and commercial development of the south rim of the Grand Canyon, at that point still a national forest property protected by the Antiquities Act (1906) as a national monument, and the other on term permit developments at Lake Mary outside of Flagstaff, Arizona. The Term Permit Act of the previous year had spurred considerable construction in the region, and foresters were necessarily drawn into the planning process. In the Lake Mary case, Leopold urged the forest service to exert stricter control over term permit siting, arguing that a portion of the lakefront should be kept undeveloped for campers. But it was another term permit case, a few years later, which would have a greater impact on the birth of wilderness policy.

In 1919, the forest service, in an unprecedented move, hired a landscape architect named Arthur Carhart to provide recreational planning for the most heavily used national forests of Region II, including several in Colorado and the Superior National Forest in Minnesota. One of Carhart's assignments was to plat term permit cabins along the shore of Trapper's Lake in Colorado's White Mountain National Forest. The relative isolation of Trapper's Lake, even today, attests to how far into the national forests Americans were taking their automobiles. Although Carhart was not against recreational development—in fact, he was an avid proponent of getting the forest service to build public facilities—he worried that the term permit system allowed for the private monopolization of the national forests' finest scenic resources, a concern not far removed from those that had motivated the protection of places such as Niagara Falls and Yellowstone. So Carhart suggested to his superiors that the shore of Trapper's Lake be saved from *all* such development, a suggestion they accepted.

Some have argued that Carhart's policy for Trapper's Lake represented the first instance of modern wilderness preservation, but such a contention misses important distinctions between his suggestion and the wilderness policy Aldo Leopold outlined a few years later.[10] In preserving national forest land both *for* recreation and *from* recreational development, Carhart's efforts at Trapper's Lake foreshadowed, and directly influenced, the concerns that Leopold would bring to wilderness policy. Indeed, in December 1919, Leopold visited Carhart to discuss the Trapper's Lake case. Leopold shared Carhart's concerns, but he had a larger vision than simply protecting "scenic territories" from private monopolization. He saw in the national forests opportunities for maintaining vast areas in a primitive state, free not only from term permit cabins but also from roads, mechanized transport, and most other forms of modern human land use. The scale of Leopold's vision was grander.

Leopold codified his thoughts on wilderness preservation in a landmark 1921 article in the *Journal of Forestry* titled "The Wilderness and Its Place in Forest Recreational Policy." Leopold began by arguing that the forest service ought to preserve some of its lands for recreation and from resource extraction, a controversial suggestion in its own right. But his chief innovation was in distinguishing wilderness from other forms of *recreational* preservation. "By wilderness," Leopold wrote, "I mean a continuous stretch of country preserved in its natural state, open to lawful hunting and fishing, big enough to absorb a two-weeks pack trip, and kept devoid of roads, artificial trails, cottages, and other works of man." "The majority," he continued, making clear his sense of the preeminent threats, "undoubtedly want all the automobile roads, summer hotels, graded trails, and other modern conveniences that we can give them. But a very substantial minority, I think, want just the opposite." What they wanted, he thought, was wilderness. Leopold ended his article by suggesting as a candidate area the headwaters of the Gila River in New Mexico's Gila National Forest.[11] By 1924, the forest service had taken his advice. With the designation of the Gila Wilderness Area, modern wilderness preservation was born.

Aldo Leopold wrote extensively about the wilderness idea during the mid-1920s, and in the process he became the nation's chief wilderness ideologue. Many of Leopold's wilderness essays were tinged with frontier romanticism, comparing the days of the covered wagon with the modern motorized era and celebrating the masculine virtues of wilderness hunting. In other pieces, he lamented the demise of uncharted places and the spiritual loss that occurred as remote corners of the earth were mapped and charted. But all of these pieces shared a concern about the impacts of roads, cars, and a new modern culture of outdoor recreation. Leopold insisted that protecting wilderness areas meant making crucial distinctions between contending recreational desires at a time of unprecedented eagerness to develop the public lands for modern recreation. It was this insistence that separated Leopold from Muir and his generation of activists, who had equated national park creation and wilderness preservation and who had not had to contend with the automobile and its implications.[12]

The mid-1920s saw a vigorous debate among foresters about the wisdom of preserving national forests as wilderness. Some objected that such a policy was elitist because it devoted huge expanses to the small percentage of visitors interested in primitive recreation. Nonetheless, William Greeley, the chief of the forest service, warmed to the idea. Although he was reticent to put any national forest territory completely off-limits to future resource use, he saw the virtue in protecting large areas from road building and permit-based recreational development. Greeley thought that wilderness preservation might help the forest service deal with the headaches that came with recreational visitation at a time when there was little money to handle recreational planning. But Greeley also made clear that whatever wilderness designations were made—and such designations were made at that point by the various district foresters—would last only as long as the timber resources of those areas were not in demand. He did not see wilderness preservation as necessarily permanent, a position that fell short of Leopold's goals. But Greeley's limited support for wilderness was not a cynical attempt to embrace a recreational mission. If anything, Greeley hoped that wilderness designation would contain the multiplying recreational claims being made on the national forests. Only by placing the modern wilderness policy within this specific interwar recreational context can we begin to make sense of the fact that the modern wilderness idea was an innovation of foresters.[13]

Forest service discussions of wilderness protection culminated in the creation, in 1929, of Regulation L-20, the agency's first formal wilderness policy. Regulation L-20 required that district foresters identify areas with wilderness potential and file a report on how they would be managed. The policy strongly discouraged roads and recreational permits in such areas, but they were not prohibited by rule—those decisions were still in the district forester's hands. Regulation L-20 also encouraged foresters to set aside "research reserves," smaller areas with unique biological attributes that were dedicated to scientific study. Finally, and tellingly, L-20 changed the name of these areas from "wilderness" areas to "primitive" areas. "Wilderness," forest service officials felt, suggested pristine nature. Some of the areas they had in mind for protection had been logged and grazed, or they would be administered in ways that belied a notion of wilderness as untouched and pure. As such, "primitive" areas, those only accessible by primitive means, seemed a better descriptor than did "wilderness." But not everyone shared the forest service's definitions of these two terms, and the question of what to call these areas would persist as wilderness policy developed over the next several decades.

The research reserves component of this policy deserves closer attention, because it speaks to the relationship between ecology and interwar wilderness advocacy. The inclusion of research reserves came in response to the lobbying of the Ecological Society of America (ESA), founded in 1915. The ESA's Committee on the Preservation of Natural Conditions, led by the eminent ecologist Victor Shelford, had been promoting "natural conditions" as a new preservationist ideal, one premised on preserving representative areas of particular ecological communities. Thus, while wilderness proponents sought large roadless areas for primitive recreation, interwar ecologists sought the preservation of small and biologically distinctive areas for scientific study. That distinction would soon be institutionalized. As wilderness advocates went on to found the Wilderness Society, the ESA effort evolved, over the course of the 1930s and 1940s, into the Nature Conservancy, an organization dedicated to protecting smaller patches of biological diversity, increasingly on private lands. If one were to look for the influence of ecology on preservationist policy during the interwar period, then, one would find the strongest evidence among the ESA's efforts to preserve natural conditions, not among efforts to preserve large-scale wilderness areas. Although Aldo Leopold did embrace a more ecologically informed wilderness ideal during the late 1930s and 1940s, such a development should not obscure the centrality of Leopold's recreational critique. Indeed, the influence of ecological thinking on the birth of modern wilderness policy was minimal.[14]

Despite the relative weakness of the forest service's first wilderness policy, advocates such as Leopold generally were happy with the result. To Leopold, getting the forest service to explicitly recognize and preserve lands within its domain that had primitive recreational values was a huge victory against the forces that posed the greatest short-term threat: road building, motorized recreation, and administrative modernization. With that victory, Leopold pulled away from wilderness politics. He had moved to Madison, Wisconsin, in 1924, worked for the National Forest Service's Forest Products Laboratory for several years, and then turned his attention to game management, a field he helped to pioneer. Leopold passed along the mantle of advocacy, though he would return to the wilderness cause during the New Deal years.

Bob Marshall

Bob Marshall stepped into Leopold's shoes. The scion of a wealthy New York family, Marshall earned a forestry degree from Syracuse University's new forestry school in 1924. He then went to work for the forest service in the northern Rocky Mountains, where he was within walking distance—he routinely hiked forty miles in a day—of some of the nation's most remote national forest backcountry. And he quickly became a vocal critic of roads that threatened to whittle away at this vast landscape.

Marshall dove into forest service wilderness discussions during the late 1920s. In the summer of 1928, he published an essay titled "The Wilderness as Minority Right," which came in response to a stinging critique by another forester, Manly Thompson. Thompson charged that the wilderness policy's true intent was to keep the "hoi polloi" out of the forests so that an elite crowd seeking primitive recreation could enjoy these areas unmolested. Defending against such charges of elitism was a constant chore for interwar wilderness advocates. Marshall rejected Thompson's characterization of wilderness as elitist, though he recognized that there were relatively few Americans who desired a wilderness experience, at least compared to those who enjoyed auto camping and motor touring. Marshall sincerely hoped that would change; indeed, he devoted much of his career to promoting accessible wilderness recreation. But he also realized that defending wilderness from roads and recreational modernization meant relying on a minority rights argument, which became a hallmark of interwar advocacy. Even if they were a minority, Marshall reasoned, those who wanted wilderness ought to be afforded areas to meet their needs. Americans had, by the late 1920s, plenty of scenic landscapes into which they could drive; the desires of the majority were well met. But the few opportunities remaining for primitive recreation were dwindling because the majority, and those who served them, continued to press for greater motorized access. Pushing roads into remaining wild areas would result, Marshall wrote, in a denial of rights to a deserving minority without substantially augmenting the rights the majority already enjoyed.

More to the point, Marshall insisted that roads and cars did more than provide access; together they fundamentally changed an area and the recreational experiences possible within it. For Marshall, wilderness was primarily a place of solitude where one went to escape the forces of modernity and experience a sense of humility, even danger. For Marshall, driving through the wilderness was an oxymoron, though convincing the public and even some policy makers to see such a fundamental tension between automobiles and wilderness was a challenge. In 1930, in one of the seminal wilderness articles of the era, "The Problem of the Wilderness," Marshall called for the "organization of spirited people who will fight for the freedom of the wilderness," not only to protect wilderness as a minority right but to build a constituency for wilderness preservation as part of a balanced public lands system. That call would bear fruit several years later.[15]

As Marshall emerged as Leopold's chief disciple, he also wrote critically about the direction of utilitarian forestry and its relationship with wilderness preservation. In the early 1930s, Marshall was one of a number of vocal critics of the forest service, a group that included Gifford Pinchot. Never did Marshall see conservation and preservation as being at odds; indeed, he would insist that better forestry practices could contribute to wilderness preservation by limiting the land base needed to meet the nation's timber needs. Marshall's example suggests that historians—and environmentalists—have drawn too sharp a contrast between utilitarian conservationists and preservationists, a contrast that has kept us from fully appreciating the critique of recreational trends that was at the core of interwar wilderness advocacy. Like Leopold, Marshall was a trained forester who remained committed to forestry practices even as he became devoted to wilderness preservation. He saw no philosophical tensions between these two commitments.

Marshall left the forest service in the late 1920s to complete a Ph.D. in botany at Johns Hopkins and to fulfill a lifelong dream of visiting Alaska. During a lengthy stay there, he began to think about the relationship between wilderness preservation and Native American subsistence economies. Marshall then spent the mid-1930s as chief forester for the Bureau of Indian Affairs, where he developed a wilderness policy for reservation lands—a policy whose goal was to protect Indian peoples and their traditional economies, not to remove them as a prelude to constructing a pristine recreational landscape. This policy was not without its naïveté and paternalism, but neither was it a policy of dispossession of the sort that had been utilized by certain advocates of national park preservation.[16] Indeed, Marshall would be a key early figure in crafting the Alaskan ideal of "inhabited wilderness," which made room for native subsistence resource use.[17] Like Leopold, Marshall was a wilderness romantic, but that romanticism was accompanied by both a thoughtful approach to the social implications of wilderness preservation and a critical look at land and resource use.

When Bob Marshall became the forest service's first director of recreation and lands in 1937, he prioritized a more permanent wilderness policy. In 1939, he realized that goal when the forest service announced its new U Regulations. The U Regulations gave the power to recommend wilderness designation to the chief of the forest service and the power to create such areas to the secretary of agriculture, moving the process up the chain of command from the district level. Moreover, they permanently prohibited all commercial timber harvests on designated areas, an important departure from L-20. Finally, the U Regulations changed "primitive" back to "wilderness" and provided not only for the creation of "wilderness areas" of more than 100,000 acres, but also for "wild areas" as small as 5,000 acres. In sum, the U Regulations crafted a stricter and more permanent wilderness system within the national forests. Fittingly, one of the first areas to be classified under the U Regulations was the Bob Marshall Wilderness Area in Montana. Two months after the U Regulations were announced, Marshall died of a heart attack at the age of thirty-eight. He willed a substantial portion of his estate to the cause of wilderness preservation, propelling the Wilderness Society into the postwar era.

Benton MacKaye

The most idiosyncratic voice for wilderness during the interwar era was that of Benton MacKaye, best known today for his 1921 vision for an Appalachian Trail (AT).[18] MacKaye too was a forester. He received a forestry degree from Harvard in 1905 and worked for the forest service on and off for the next decade. During that time, he distinguished himself not as a wilderness advocate but as a critic of the service's failure to grapple with the social aspects of forestry. The timber industry was rife with labor unrest, and MacKaye argued that the forest service needed to support just and sustainable timber communities as well as resource sustainability. But the forest service was not interested in his social forestry, and so MacKaye moved to the Labor Department in the mid-1910s. His efforts there culminated in a 1919 government report, *Employment and Natural Resources*, in which he argued for the creation of socialist resource communities on the public lands to accommodate returning soldiers and the unemployed. Two years later, MacKaye repackaged these ideas in his AT proposal, in which he envisioned a series of resource communities strung along the Appalachian chain and connected by a recreational trail. With its emphasis on community and resource development, MacKaye's AT proposal was not a wilderness vision.

MacKaye became a wilderness advocate as he defended and redefined the AT in the automobile age. In the early 1920s, MacKaye fell in with a group of urban and regional planners, among them Lewis Mumford, with whom he helped to found the Regional Planning Association of America (RPAA) in 1923. It was in this context that MacKaye reconceptualized the AT as an open space barrier set against the spread of what he called "metropolitanism"—what we would today call "sprawl." By the end of the decade, even the AT, distant as it was from America's urban centers, faced the threat of metropolitanism. In particular, MacKaye opposed a series of New Deal recreational parkways, including the Skyline Drive in Shenandoah National Park and the Blue Ridge Parkway. These threats drove his social vision of the trail to the background. By the early 1930s, MacKaye became convinced of the need to defend the AT as a "wilderness trail," by which he meant a trail far from automobiles, modern roads, and other sights and sounds of modernity. MacKaye and trail activists Harvey Broome and Harold Anderson were preparing to found an advocacy group for just such a purpose when Bob Marshall crossed their path and suggested an organization with a broader scope.

The Wilderness Society was the immediate product of this collision between New Deal recreational roads and a trail that MacKaye came to define in wilderness terms. But MacKaye's vision was not entirely a defensive one. Indeed, as a regional planner, he brought to interwar wilderness advocacy a grand vision of what one scholar has called "the city-shaping possibilities of open space conservation."[19] Not content to see wilderness solely as a place apart, MacKaye continued during his long life (he was ninety-six when he died in 1975) to envision trails and various other "wilderness ways" as tools for reinjecting wilderness into modern landscapes. To the extent that wilderness activists and planners have embraced greenways and connectivity as planning goals, they are inheritors of MacKaye's unique contribution to modern wilderness advocacy.

Robert Sterling Yard

Robert Sterling Yard came to wilderness advocacy from the national parks establishment.[20] His close friend Stephen Mather hired him in 1915 to publicize and help to build a constituency for the parks—a job that Yard performed with an effectiveness he would later regret. With Yard's help, Mather steered through Congress the Organic Act of 1916, which created the National Park Service and its now-famous dual mandate both to promote the enjoyment of the parks and to protect park resources for future generations. In 1916, there seemed little reason to believe that those mandates would conflict.

Yard left the park service to help create the nongovernmental National Parks Association (today the National Parks Conservation Association) in 1919, and he ran its daily affairs for more than a decade. In that capacity, he developed a two-pronged commitment to the "complete conservation" of the national parks. First, he insisted that designated national parks were not to be developed for their natural resources, though he initially saw little problem with tourist developments. Second, Yard doggedly defended the scenic standards of the national park system, insisting that only the most monumental or sublime areas ought to be national parks. In this sense, he was like a preservationist of Muir's generation. In protecting park standards, Yard often opposed proposed parks when he thought their scenery substandard, which was not usually a popular position among park service officials. Indeed, during the 1920s, Yard went from being a publicist to a watchdog whose notions of what constituted a proper national park were quite conservative.

Ironically, that conservatism would steer him in a radical direction. Yard grew concerned, as the 1920s progressed, by how wedded to motor tourism and tourist development the park service had become. In the process, he came to several important realizations. First, he argued that tourist development within the parks constituted an invasion akin to resource development. Second, he railed against pork barrel park politics, a process whereby politicians and booster groups urged new—and usually, by his tastes, substandard—national parks so as to profit from the government and tourist dollars that would follow. Third, he concluded that his own prodigious advertising efforts had been folded into a brand of mass nature tourism that was undermining what he saw as the high cultural purposes of the parks. And so Yard became a wilderness advocate as a way of calling the bluff of national park promoters and developers. If park service officials, politicians, and boosters were truly interested in preservation, and not merely in the development of parks as lucrative tourist destinations, they would preserve them largely as wilderness areas, he argued, without roads and modern tourist facilities.

By the early 1930s, Yard's thorny activism had thoroughly frustrated most other park advocates. Increasingly, he found his home among the wilderness activists coming out of forest service circles. In Yard's hands, the modern wilderness idea was born of a strong critique of the park service's model of preservation as tourist development. Yard's story suggests that, during the interwar era, tensions between wilderness advocates and national park administrators were more potent than conflicts between preservationists and resource developers.

Yard's park advocacy was not completely futile, particularly once he had the clout of the Wilderness Society behind him. At the end of the interwar era, the park service created several "wilderness" national parks, such as Olympic and King's Canyon, that were largely free of roads, while it carefully scrutinized road building in others, such as the Great Smoky Mountains National Park and Everglades National Park. Nonetheless, the park service remained a reluctant designator of wilderness within its own bounds well into the postwar years.

Conclusion

In the years between the world wars, wilderness advocacy and policy grew not merely as a progressive intellectual development, as traditional narratives of wilderness history would have it, but in reaction to a specific set of forces—the automobile, roads, and modern recreational trends—new to the era. In this context, the modern wilderness idea stood as a critique not only of the consumerism driving interwar outdoor recreation but also of the ways in which automobility changed the American landscape and American culture. Interwar wilderness thinking stood in stark contrast to the wilderness thought of the late nineteenth and early twentieth centuries, when the idea had been deployed by national parks advocates to counter resource exploitation of various sorts. That the modern wilderness idea was largely the invention of committed utilitarian foresters, several of whom were quite radical, suggests the unusual contours of interwar environmental politics.

In the postwar era, the context changed yet again, as discussed by Mark Harvey in the following chapter. Threats posed to wilderness by the development of public land resources reemerged, making it easy to forget the forces that had produced the modern wilderness idea in the first place. One should not exaggerate this shift, as there were strong continuities from the interwar era. Road building and automobile use continued to fuel an outdoor recreation boom that, in terms of scale, dwarfed what had happened during the interwar years. But, even for the most committed wilderness activists, such concerns took a back seat to two trends that altered postwar wilderness debates: dam building in the undeveloped canyon country of the West and the movement of industrial timber production into the national forests.[21] Together, these threats reoriented wilderness politics around debates between resource users and the conservation bureaus that served them, on the one hand, and preservationist groups, which grew to rely on the very recreational constituencies whose behavior interwar activists had critiqued, on the other. Indeed, these debates—over places such as Echo Park, Glen Canyon, and the ancient forests of the Pacific Northwest—seemed such a direct echo of the Hetch Hetchy controversy that historians have assumed a continuity between the Progressive and postwar eras, overlooking the very different interwar context that gave birth to modern wilderness advocacy. Only since the 1990s, with the rise of a new set of questions about motorized access to the nation's public lands, have wilderness politics returned to a context that would be recognizable to these interwar advocates.

Notes

1. This chapter is drawn from my book, *Driven Wild: How the Fight against Automobiles Launched the Modern Wilderness Movement* (Seattle: University of Washington Press, 2002).

2. For the preeminent example of the traditional account, see Roderick Nash, *Wilderness and the American Mind*, 3d ed. (New Haven, Conn.: Yale University Press, 1982; originally published in 1967).

3. *Historical Statistics of the United States, Colonial Times to 1970*, part 2 (Washington, D.C.: GPO, 1975), 716.

4. Warren James Belasco, *Americans on the Road: From Autocamp to Motel, 1915–1945* (Cambridge, Mass.: MIT Press, 1979).

5. Marguerite Shaffer, *See America First: Tourism and National Identity, 1880–1940* (Washington, D.C.: Smithsonian Institution Press, 2001).

6. Visitation statistics are from the National Park Service's Web site: http://www2.nature.nps.gov/stats/decademain.htm.

7. *Proceedings of the National Conference on Outdoor Recreation*, 68th Cong., 1st sess., May 1924, S. Doc. 151; *Proceedings of the Second National Conference on Outdoor Recreation*, 69th Cong., 1st sess., 1926, S. Doc. 117.

8. *Living Wilderness* (September 1935), 1.

9. Curt Meine, *Aldo Leopold: His Life and Work* (Madison: University of Wisconsin Press, 1988).

10. Donald Baldwin, *The Quiet Revolution: The Grass Roots of Today's Wilderness Preservation Movement* (Boulder, Colo.: Pruett, 1972).

11. Aldo Leopold, "The Wilderness and Its Place in Forest Recreational Policy," *Journal of Forestry* 19, no. 7 (November 1921): 718–21.

12. Most of Aldo Leopold's wilderness writings are collected in *The River of the Mother of God and Other Essays by Aldo Leopold*, ed. J. Baird Callicott and Susan Flader (Madison: University of Wisconsin Press, 1991).

13. Manly Thompson, "A Call from the Wilds," *Service Bulletin* (May 14, 1928): 2–3.

14. Victor Shelford, "Preserves of Natural Conditions," *Transactions of the Illinois Academy of Science* 13 (1920): 37–58; Shelford, ed., *Naturalist Guide to the Americas* (Baltimore, Md.: Williams and Wilkins, 1926); Robert Croker, *Pioneer Ecologist: The Life and Work of Victor Ernest Shelford, 1877–1968* (Washington, D.C.: Smithsonian Institution Press, 1991).

15. Bob Marshall, "The Wilderness as Minority Right," *Service Bulletin* (August 27, 1928), 5–6; Marshall, "The Problem of the Wilderness," *Scientific Monthly* 30, no. 2 (February 1930): 141–48.

16. See Mark David Spence, *Dispossessing the Wilderness: Indian Removal and the Making of the National Parks* (New York: Oxford University Press, 1999); Karl Jacoby, *Crimes against Nature: Squatters, Poachers, Thieves, and the Hidden History of American Conservation* (Berkeley: University of California Press, 2001).

17. See Theodore Catton, *Inhabited Wilderness: Indians, Eskimos, and the National Parks in Alaska* (Albuquerque: University of New Mexico Press, 1997).

18. Benton MacKaye, "An Appalachian Trail: A Project in Regional Planning," *Journal of the American Institute of Architects* 9 (October 1921): 325–30. See also Larry Anderson, *Benton MacKaye: Conservationist, Planner, and Creator of the Appalachian Trail* (Baltimore, Md.: Johns Hopkins University Press, 2002).

19. Kermit Parsons, "Collaborative Genius: The Regional Planning Association of America," *Journal of the American Planning Association* 60, no. 4 (Autumn 1994): 466.

20. See Robert Shankland, *Steve Mather of the National Parks*, 3d ed. (New York: Knopf, 1970); John Miles, *Guardians of the Parks: A History of the National Parks and*

Conservation Association (Washington, D.C.: Taylor & Francis, 1995); Shaffer, *See America First.*

21. See Paul Hirt, *A Conspiracy of Optimism: Management of the National Forests since World War Two* (Lincoln: University of Nebraska Press, 1994); Mark Harvey, *A Symbol of Wilderness: Echo Park and the American Conservation Movement* (Albuquerque: University of New Mexico Press, 1994).

Eleven

Loving the Wild in Postwar America

Mark Harvey

The most striking development in Americans' relationship with wilderness following World War II was its soaring popularity. Bolstered by the nation's triumph in the war and ready to celebrate, Americans took to the nation's roadways and eagerly made their way to places where they saw wildlife in their natural habitats, along with geysers and boiling mud pots, deep canyons, giant waterfalls, and snow-capped peaks. For many Americans and travelers from abroad, such spectacles provided an enticing taste of wilderness with a minimal exertion of effort, often from an automobile or from a vantage only a few steps away from the road. Others, though, found their wilderness far removed from cars and roadways, typically in the backcountry of national parks or forests, where spectacular animals, plants, and scenery less often were seen. While people found wilderness in different locales, all seemed to view it as a place where nature was in its purest form and where the contrast with urban, suburban, and rural landscapes was starkly clear. As growing numbers also realized that wilderness lands, despite their great popularity, were at risk of being reduced in size or ecological integrity because of weak laws governing their protection, a movement emerged to establish a national wilderness system, which culminated in passage of the Wilderness Act of 1964.

Building a Wilderness Movement by Experiencing Wilderness

The nation's growing appetite for wilderness from the 1940s to the 1970s built on the long American fascination with wilderness described in this book. This desire to see and experience wilderness was fed by a rapid expansion of outdoor recreation following World War II, sparked by cheap unrationed gasoline, higher living standards, and paid vacations. The American automobile culture, a key force in shaping the burgeoning suburbs and a symbol of rising prosperity, ensured that millions of people could reach the wilderness that ever more of them craved. Along with the Cold War and national security concerns, the surge in automobile ownership sparked passage of the Interstate Highway Act of 1956, a law of monumental importance in spurring travel and a rising tide of visitors to state and national parks, monuments, historic sites, and national forests.[1]

As explained by Paul Sutter in the previous chapter, in the 1920s and 1930s, wilderness activists had lobbied to preserve places without roads, motorized vehicles, and the sounds of mechanical civilization. Defenders of wilderness in the postwar years continued working to set aside roadless areas, yet they also recognized that without automobiles, motor homes, and other vehicles, travelers would likely not reach the wilderness in the first place. Americans now enjoyed the wilderness from their vehicles, taking in stunning views of the Grand Canyon, the Tetons, or the steep granite walls of Yosemite Valley from roadside turnouts. Through this "windshield experience," millions of Americans had a taste of the nation's scenic spectacles that they identified as "wild."[2]

In the postwar years, a growing number of middle-class families oriented their summer vacations around a circuit of national parks. Yellowstone, Glacier, Mount Rainier, and other parks now became "a must see" with the traveling public, and many took it as an obligation of citizenship to gaze upon the depths of the Grand Canyon, imposing redwoods, or the Lower Falls of the Yellowstone River. Stories about the lengths to which some individuals went to fulfill this obligation are revealing. For example, one small-town midwestern barber drove 1,500 miles to Yellowstone National Park, watched Old Faithful spout, then returned to his car and drove home to Wisconsin without stopping at any other historic or scenic spot. To him, Old Faithful was a nationally scenic icon that offered a satisfying taste of wild nature. Such was the allure of America's wildlands in the postwar years.

Some people found a deeper, more intimate experience in their encounters with particular wilderness landscapes. For them, wild areas were special places visited by those able and willing to expend the physical energy and display the endurance that were often required. For hunters, anglers, day hikers, backpackers, rock climbers, and photographers, the backcountry of national parks, a few state parks like the Adirondacks, and the roadless areas in national forests provided unique natural settings that drew them irresistibly. To anglers and hunters from Idaho and Montana, the Selway-Bitterroot Primitive Area spanning the border of those states was a weekend haven; to hikers and backpackers in Oregon, the Three Sisters Wilderness Area in the Central Cascades proved an alluring destination. Still others flocked to the northern Rockies where the Bob Marshall Wilderness Area or the vast backcountry of Glacier National Park provided ample opportunities for getting into the wild. Their experiences often sparked efforts to defend these areas from loggers, miners, grazing livestock, and motorized vehicle users. Recreational wilderness users formed the backbone of an organized movement that became a vibrant part of the postwar conservation establishment.

Expanding travel and tourism gave rise to a mass audience hungry for stories and images of wildlands, which appeared in books, newspapers, films, and the publications of conservation organizations. Popular magazines, such as *National Geographic*, *Sunset*, and *Arizona Highways*, provided a stream of captivating wilderness images, while television productions and films further fed the nation's appetite for wildlife and scenery. The Walt Disney Corporation produced a series of true-life adventure films, with *Seal Island*, *The Olympic Elk*, *The Living Desert*, and *The Vanishing Prairie* among the most popular. A specialized genre of books, magazines, and films catered to the more adventurous. Wildlife photographers Herb and Lois Crisler produced films with special appeal to hikers and backpackers, including *White Wilderness*, devoted to wolves in Alaska, and *The Living Wilderness*, which highlighted the wildlife and rainforests of Olympic National Park in Washington. Olaus and Mardy Murie, leading figures in the Wilderness Society, narrated *Letter from the Brooks Range*, a film chronicling their experiences from the summer of 1956 on the Sheenjek River in northern Alaska. Mardy Murie's book *Two in the Far North* later became a classic narrative of wilderness adventure.[3]

In the late 1960s and 1970s, the books of Edward Abbey, most notably *Desert Solitaire* and *The Journey Home*, reached a younger generation of wilderness enthusiasts. In *Desert Solitaire*, Abbey insisted that a genuine wilderness experience was impossible so long as one was anywhere near an automobile. "Do not jump into your automobile next June and rush out to the canyon country hoping to see some of that which I have attempted to evoke in these pages," Abbey cautioned. "In the first place you can't see *anything* from a car; you've got to get out of the goddamned contraption and walk, better yet crawl, on hands and knees, over the sandstone and through the thornbush and cactus."[4]

Films, printed images, and richly drawn prose portrayed "wild" nature bereft of people or obvious marks of human habitation, appealing places that stood in sharp contrast to the suburban and urban landscapes where Americans conducted their everyday routines. Wilderness was defined as nature in its purest state, unaltered by human hands, unadulterated by industry or agriculture, and seemingly untouched through eons—the wilderness ideal seen in many of the preceding chapters of this book. In this sense, wilderness was a state of mind as much as a specific place, an original American landscape defined by its exotic qualities of wildness.[5]

If rising prosperity and greater leisure time helped to spawn the postwar throng of wilderness enthusiasts, improvements in outdoor equipment helped them get into the backcountry. The rucksack used by day hikers, for example, had been a bulky, awkward, and primitive piece of equipment, hard on the shoulders and back. By the 1960s, firms like Recreational Equipment Incorporated (REI) and Kelty constructed lightweight aluminum frames which supported water-resistant nylon and thick padding, resulting in a backpack that rested on the hips rather than the shoulders, which could be worn comfortably for many hours. The new backpacks, along with an array of lightweight nylon tents, sleeping pads, freeze-dried foods, and cooking equipment, enabled a growing number of hikers to spend several days and nights in remote locations.[6]

For many outdoors enthusiasts, a thrilling sight or episode became a defining moment of their encounters with wilderness, such as watching grizzly bears roaming freely in the backcountry of Glacier or Yellowstone national parks or marveling at richly colored meadows of wildflowers in the Rockies, home to sky pilots, columbine, and Indian paintbrush. In wilderness, one experienced summer blizzards in mountainous terrain, or thunderstorms punctuated by torrential downpours and sizable hail, followed by dazzling bright sunshine glinting from tall pines and mirrored in crystal-clear streams. Wilderness lovers frequently encountered others along the trail or at canoe portages who had experienced similar moments, and in the shared stories found themselves part of a community. Though not necessarily bound by political allegiances or economic status, they shared a deep love of the wild.

The link between physical activity and wilderness appreciation remained important in the postwar years, as it had been for so many earlier wilderness advocates, from Thoreau to the men and women of the Appalachian Mountain Club. Some activists who fought to protect wildlands became so engaged because of the pleasures and satisfaction resulting from the rigorous physical exertion they had expended in reaching wilderness areas. On occasion, wilderness defenders put their enjoyment of physical activity on display to help publicize threatened areas they hoped to protect. In 1954, Supreme Court justice William O. Douglas led hikers on an eight-day walk along the 200-mile towpath of the old C&O Canal between Cumberland, Maryland, and Washington, D.C., which was threatened with being paved over for a highway. Douglas helped to launch the widely publicized event after the *Washington Post* published his guest editorial, which advocated saving the towpath and canal. Among the several dozen hikers accompanying him were editorial writers of the *Post*, who later changed the paper's position and called for the canal to be preserved. Four years later, Douglas again joined a few dozen others and staged a similar hike along a stretch of Pacific Coast beach in the state of Washington that was also threatened by a highway project.[7]

People's individual experiences—whether walking in the woods or looking through the windshields of their cars—proved crucial in forging a politically active community of wilderness enthusiasts. Many joined organizations which touted the beauties and appealing flora and fauna of their preferred areas. "Friends of the Wilderness" organizations relied on newsletters and other publicity to draw attention to wildlands around the country, such as the Minnesota-Ontario Boundary Waters, Three Sisters and Glacier Peak wilderness areas, and Adirondacks Forest Preserve. Members donated time and money to help maintain and protect the places they loved.

In the parlance of the time, wilderness was considered (and promoted as) "pristine" or "virgin" land, yet such labels often masked more complicated landscape histories and ecologies. In observing these areas firsthand, wilderness lovers often found evidence of mining, logging, grazing, and road construction. Owing to the Mining Law of 1872, which permitted claims to be patented (allowing prospectors to file new claims and convert them into private land), some forest service primitive areas had been pockmarked with mining claims and jeep trails. Elsewhere, overgrazing took a heavy toll on water quality and rangelands; the Gila Primitive Area in New Mexico, the first area set aside by the forest service as a wilderness in 1924, comprised numerous areas damaged by heavy grazing.[8] The discovery of roads, mining claims, and overgrazed range inside parks and primitive areas discouraged wilderness proponents, but also bolstered their interest in strengthening the hands of the managers administering the areas. The disjuncture between the ideal of a pristine wilderness and the reality of used (and sometimes abused) landscapes created activists.

Scientists based at universities conducted research on the geology, forest ecology, and animal behavior within wilderness areas. The knowledge they accumulated provided scientific arguments for establishing stronger regulations and laws protecting the lands, waters, and species of wilderness. Scientific arguments to promote conservation had been seen ever since G. P. Marsh (see chapter 4); his evidence correlating deforestation with degraded watersheds and soils had been employed to justify the creation of national forests and the Adirondack Forest Preserve in the late nineteenth century. Postwar wilderness proponents advanced similar arguments to defend protected areas against dams, logging, and overgrazing. Bernard Frank, a hydrologist with the forest service and a founding member of the Wilderness Society, published articles linking watershed quality to the protection of primitive areas in national forests in the West.[9] Frank and other hydrologists also lent their expertise to conservationists seeking to thwart logging and dam projects in the Adirondack Forest Preserve. Friends of the Adirondacks applauded the arguments that linked a reliable water supply for New York City with the need to uphold the "forever wild" clause of the New York state constitution. They understood how New York lawmakers, who felt pressured to permit more logging and water projects within the Adirondacks, could resist those efforts by emphasizing that keeping the Adirondacks "forever wild" helped to ensure water to urban residents downstate. In just this way, urban dwellers in the East and West became part of the constituency for wilderness protection.[10]

Still, although ecological ideas were valuable, wilderness proponents commonly relied on the social and cultural arguments developed during previous eras of American interactions with wilderness, particularly those of the Progressive Era. The postwar generation witnessed dramatic changes to the landscape from the sprawl of suburbs and from increased logging and dam construction. As a result, they emphasized the many benefits of places that humans chose not to disturb, where adventure, spiritual uplift, and valuable knowledge might be gained, as well as the opportunity to encounter the challenges of frontier life which had shaped the American character.[11] Sigurd Olson of Ely, Minnesota, a veteran canoeist of the Boundary Waters and a committed activist in their protection, thrilled in following the same canoe routes and portages as Alexander MacKenzie, Sieur de la Verendrye, and other fur traders from centuries past. He was captivated with their journals, and he adored the names that they left: Lac La Croix, Deau Riviere, Saganaga, and Kahnipiminanikok. "When I entered the fastnesses of the Quetico-Superior I would become a part of all that," he wrote. "It would be like lifting the curtain on another world. No longer would I belong to the twentieth century. I would be a voyageur of the seventeenth, a man from Trois Riviere or Montreal, I would see the country through his eyes."[12] For Olson, to enter the wilderness meant taking a journey into the past.

Most wilderness sojourners experienced a sense of awe in the Boundary Waters, beneath tall mountains, or in the midst of the canyonlands, and many wrote about it in poetry and prose.[13] Howard Zahniser, executive secretary of the Wilderness Society, expressed his awe of wild nature following a horsepack trip into the Cloud Peak Primitive Area in Wyoming's Big Horn range in 1947:

> Constantly disintegrating, attacked by hail and rain and lightning, crumbled by frost and ice and burdened with snow, the mountain time and again had yielded a part of itself to the always tugging pull of the earth's gravity. Yet there it still stood, the debris of the elements all about, itself an aspect of awe. One could no more see at the moment the destruction of this mountain than he could perceive its whirling with himself through space. One could no more imagine the final passing of this mountain than he could anticipate his own disintegration. Yet somehow or other in this Presence, this bright Sabbath noon, one found himself reflecting on both and admiring and coveting this inert mountain's aspect of noble serenity in these eternal processes of dissolution.[14]

The passage contains more than a hint of religious language and evokes a sense of John Muir's belief that one finds God in the high mountains. The language and imagery, like much of Muir's, is transcendent, drawing attention to vast and powerful forces in the earth and to the Creator.[15] Influenced by his father and four uncles, each of them ministers, Zahniser (yet another minister's son bridging God and wilderness) embraced an ethic of stewardship and preached that humans had a duty to protect wildlands. "To know the wilderness is to know a profound humility," he said in a speech in 1955, "to recognize one's littleness, to sense dependence and interdependence, indebtedness, and responsibility. Perhaps, indeed, this is the distinctive ministration of wilderness to modern man."[16]

Threats to the Wilderness

For Zahniser and others, the growing affluence of the postwar years was paradoxical. On one hand, the steady growth of tourists and outdoor lovers helped to build the membership bases of the Wilderness Society, the Sierra Club, and other organizations. On the other hand, the recreational explosion brought forth many others who relished the speed and thrills of motorboats and off-road vehicles—a different type of sublime experience. The resulting faultline gave rise to several key battles. The most protracted of these centered on the Boundary Waters area in Minnesota and northwestern Ontario, Canada, where private landowners and anglers used airplanes to access prime fishing areas within forest service roadless areas. Taking umbrage at the intrusions of planes and motorboats, wilderness proponents raised funds to acquire private lands and called for tighter restrictions on the use of motors. Late in 1949, President Harry S. Truman proclaimed an air space reserve, which prohibited planes below 4,000 feet over the roadless areas. Truman's proclamation strengthened protection of the Boundary Waters and greatly encouraged wilderness activists nationwide.[17] His decision, of course, applied solely to the Boundary Waters, and in subsequent years conflicts over all-terrain vehicles, motorboats, jet skis, and snowmobiles surfaced in many wilderness areas and national parks.

Other powerful threats to wilderness in the postwar years came from natural resource industries and from government agencies concerned with water development. The rapidly expanding housing market, baby boom, and growth in manufacturing during the postwar economic expansion combined to bring about increased mineral extraction and timber harvesting, along with attempts to open new fields of oil and natural gas on public lands. Such pressures gained strength from the Cold War, which placed a premium on the full development of minerals, timber, fossil fuels, and hydroelectric power deemed to be crucial to the nation's economy and security. Wilderness advocates were dismayed by logging companies that attempted to reduce the size of Olympic National Park to make available more land for timber harvesting and were disturbed by attempts of oil and gas firms to gain access into forest service primitive areas.

They were also disheartened by proposals to dam rivers within or near primitive areas or portions of the national park system. From the middle to the late 1940s, the Black River Regulating District Board proposed a Higley Mountain dam and reservoir along the south branch of the Moose River flowing westward out of the central Adirondacks. At that same time, the Bureau of Reclamation sought to dam the outlet stream of Lake Solitude in the Cloud Peak Primitive Area of the Big Horn range, and the Army Corps of Engineers sought to build a dam on the north fork of the Flathead River bordering Glacier National Park in Montana.[18]

The biggest dam controversy, though, centered on little-known Dinosaur National Monument, a huge preserve spanning the border of Utah and Colorado, which protected a steep wall of dinosaur fossils and the magnificent Yampa and Lodore canyons. In the center of the monument was Echo Park, a small valley surrounded by imposing cliffs where the Bureau of Reclamation hoped to erect a large dam in the 1950s. Having gained approval from two secretaries of the interior in the Truman and Eisenhower administrations, the dam enjoyed strong backing in the surrounding states, which coveted its water storage and hydroelectric power possibilities.[19]

Led by the Sierra Club, the National Parks Association, and the Wilderness Society, a coalition of organizations spent six years mounting a campaign to block the dam in Congress. Still relatively small and politically weak, these groups forged an alliance with water and power interests in California and with agricultural interests in the Midwest and South, which felt anxious about placing more acreage under irrigation and its potential to reduce commodity prices. Conservationists argued that the Echo Park dam would severely compromise a magnificent scenic preserve holding high wilderness values and would violate the National Park Service Act of 1916, which mandated that the parks be left unimpaired. Located in a remote corner of the West and accessible only by river or gravel road, Echo Park had rarely been visited. Dominated by a huge 800-foot-high monolith called Steamboat Rock with its sheer east face soaring over the confluence of the Green and Yampa rivers, Echo Park seemed enormously fragile with the prospect of a giant dam and reservoir in its midst.

Aware that it was little known, conservationists highlighted Echo Park in their periodicals and encouraged coverage of the controversy in magazines and newspapers. Two films—Charles Eggert's *Wilderness River Trail* and David Brower's *Two Yosemites*—shone a spotlight on the threatened monument. Brower also spearheaded publication of *This Is Dinosaur*, edited by Wallace Stegner and published by Alfred A. Knopf, to increase awareness of the historic, scenic, and wilderness values of Dinosaur National Monument and Echo Park.[20] By means of such publicity, those determined to stop the dam turned the remote canyon into a potent symbol of the nation's threatened wilderness. Pressure grew on Congress to remove the dam from the larger project.

Early in 1956, Congress did so when it passed the Colorado River Storage Project Act. The law authorized several large dams but guaranteed that no dam would be built within the national park system. The outcome marked an important triumph for wilderness advocates. A weighty coalition of organizations had been solidly welded during the Echo Park fight and gained considerable stature in the political arena and national press. Brower, the young, aggressive director of the Sierra Club, earned plaudits for his bold confrontation of the Bureau of Reclamation over the evaporation rates of its reservoirs. With the triumph over Echo Park dam, Brower helped to transform the Sierra Club from a hiking and climbing club into an influential national voice for wilderness. Tall, wiry, and with a reputation as an excellent rock climber, Brower became a central figure in the postwar wilderness movement and a symbol of that movement's growing power and influence on the West Coast. Capitalizing on the outcome of the Echo Park battle, Brower led efforts to protect Rainbow Bridge from the rising waters of Lake Powell and another high-profile campaign that stopped the Bureau of Reclamation's proposed dams in Marble and Bridge canyons within the Grand Canyon.[21]

Brower's special genius was in finding ways to heighten public awareness of the nation's great wilderness. His background as an editor at the University of California Press made him especially aware of the value of high-quality books for capturing public attention. Encouraged by the success of *This Is Dinosaur* in helping to defeat the Echo Park dam, Brower initiated a publishing program in the Sierra Club that featured coffeetable-sized works offering poetry, inviting prose, and stunning photographs. Among the best of the "exhibit format" books were *This Is the American Earth* (1960), *In Wildness Is the Preservation of the World* (1962), *Not Man Apart* (1965), *The Place No One Knew* (1963), *Time and the River Flowing* (1964), *The Last Redwoods* (1963), and *Gentle Wilderness: The Sierra Nevada* (1964). Featuring impressive color photographs by Eliot Porter, Philip Hyde, and Richard Kauffman and poetry by Robinson Jeffers, Henry David Thoreau, and others, the exhibit format books captured the attention of an affluent and influential audience.

Because of the books' high cost, Brower had to devote enormous energy to finding grants and donations to underwrite the series, and he had to defend the book program from Sierra Club board members who questioned its expense and value. Brower asserted that investment in them paid off because the books appeared on the coffeetables of well-heeled Sierra Club members in the Bay Area and across the country, and yielded a bigger club membership, greater income, and political support for wilderness.[22] Brower found an ally in board member Ansel Adams, whose mostly black-and-white photographs also nourished the public appetite for wilderness images. Adams's photographs of the Grand Canyon, Teton Range, Yosemite Valley, and Sierra Nevada were the most famous images of these places and conveyed a fresh conception of wilderness that emphasized its stark, sublime beauty. Together with the exhibit format series, Adams's photographs helped to turn the Grand Canyon, Yosemite, the redwoods, and Sierra Nevada into icons of America's wilderness.[23]

Such lavish publicity of great scenic landscapes found a ready market among the burgeoning middle class of travelers and sightseers, many of whom cherished the national parks. Yet for many involved with defending wilderness in the early postwar years, the little-known roadless and primitive areas within the national forests were much harder to bring to the public's attention, a problem that compounded the challenge they faced in pressing the forest service to protect these areas. Following World War II, rising demands for timber, grazing, and other uses generated enormous pressure on the agency to relax its wilderness regulations (the U Regulations adopted in 1939, which were discussed in the previous chapter) and to open up primitive areas to such economic uses.

From the 1940s until the early 1960s, under pressure by those eager to defend the primitive areas from encroachment, the forest service was slowly reclassifying its primitive areas into "wild" areas (5,000 to 100,000 acres) and "wilderness" areas (more than 100,000 acres). Reclassification involved fixing solid boundary lines around primitive areas set aside with little public input in the 1920s and 1930s under the earlier L-20 Regulations. Hoping that its wild and wilderness areas would reflect current commodity and recreational demands, the National Forest Service conducted public hearings in advance of the reclassifications. When some reclassifications angered wilderness advocates, the latter took action. In one of the most important of these decisions, the forest service in 1957 announced a newly classified wilderness area in the Oregon Cascades, which eliminated 53,000 acres from the Three Sisters Primitive Area. The announcement shocked wilderness activists in the Northwest, who quickly notified national leaders, including Howard Zahniser of the Wilderness Society.

A National Wilderness Act

For Zahniser, the Three Sisters decision underscored the importance of the campaign he had launched in 1956 to establish a national wilderness system that would safeguard wilderness on public lands by federal statute.[24] Only in this way, he believed, would wilderness have permanent protection. The wilderness bill proposed to give statutory protection to wild or wilderness areas, which totaled less than 10 million acres in the 1950s. The remaining primitive areas and other acreage in national parks and federal wildlife refuges could be added to the system upon further review. In short, the legislation would establish a process for adding areas to the national wilderness system, subject to congressional approval.

From 1956 until 1964, the Wilderness Society, the Sierra Club, and dozens of organizations spearheaded a campaign to enact the wilderness bill. The legislation underwent careful scrutiny by subcommittees of the House and Senate, was the focus of nineteen public hearings in Washington, D.C., and in several western cities, and generated enormous debate in newspapers and magazines.[25] Ranching, mining, and timber industries, fearing sharp reductions in their commodity production, opposed the legislation, while pointing out that minerals, timber, and grazing lands had both economic and strategic value to the nation during the Cold War. Any "locking up" of these resources, they insisted, was risky and even un-American. Officials from western states feared that more wilderness would mean diminished revenues to counties, revenues which the forest service had provided to local governments to compensate them for lands no longer subject to local or state taxation. At first, the National Forest Service and National Park Service opposed the wilderness bill, anxious that their authority over lands they administered would be curtailed.[26]

Supporters of the bill encountered another daunting obstacle in the federal tax code and Internal Revenue Service regulations, which prohibited nonprofit organizations from engaging in any "substantial" lobbying of Congress. Since the regulations did not clearly define "substantial," conservationists felt considerable uncertainty over how much they could lobby and constantly feared that their efforts to promote passage of the bill (such as visiting with lawmakers on Capitol Hill) would compel the IRS to revoke the tax-deductible status on which they relied for building membership. Zahniser, who played the central role in building a coalition in support of the bill, at times felt overwhelmed by the prospect of losing the Wilderness Society's tax-deductible status. "Enactment of the bill is so nearly possible in this present session of Congress that it would almost make you cry to have to do anything else but work on this," he wrote to a friend in 1958. "I wish I could have spent the past two months working on the Wilderness Bill that I have spent in writing or worrying about what we could or should do about it, or not do, or shouldn't do."[27]

Despite these myriad obstacles, conservationists had several factors on their side. First, the coalition formed during the Echo Park campaign held together during the battle over the wilderness bill. Its leaders included Brower, whose aggressive demeanor stood in contrast to the soft-spoken but persistent Zahniser.

Grassroots leaders across the country, representing hikers, scientists, photographers, hunters, and anglers, provided crucial support. Prominent among them were Polly Dyer of Seattle, Washington; Charlotte Mauk of Berkeley, California; Karl Onthank of Eugene, Oregon; Elliott Barker of Albuquerque, New Mexico; and Paul Schaefer of Schenectady, New York. The older hiking clubs in the East, instrumental in constructing the Appalachian Trail, provided vital support as did the sizable National Wildlife Federation and Izaak Walton League. The General Federation of Women's Clubs proved instrumental to the campaign's success as well, just as it had in the battle to save Echo Park. The women's clubs had long been a major champion of the national park system, identifying the parks with Americans' love of the land and patriotic spirit (see chapter 9 in this volume). Now they brought that perspective, and a large membership, to bear on the challenge of saving the shrinking wilderness. Publicity of wilderness in *Harper's*, *Atlantic*, *Sports Illustrated*, *Life*, and the *Saturday Evening Post* proved invaluable in reaching diverse and national audiences.[28]

Eloquent statements helped as well. Perhaps the most eloquent of all came from the pen of Wallace Stegner, author of numerous works of fiction, biographer of John Wesley Powell, and editor of *This Is Dinosaur*. Late in 1960, Stegner sent a four-page, single-spaced letter to the Outdoor Recreational Resources Review Commission, offering a powerful statement of the values of wilderness. His "wilderness letter" revealed his adoration of the Robbers' Roost country in Utah, which he called

> a lovely and terrible wilderness, such a wilderness as Christ and the prophets went out into; harshly and beautifully colored, broken and worn until its bones are exposed, its great sky without a smudge or taint from Technocracy, and in hidden corners and pockets under its cliffs the sudden poetry of springs. Save a piece of country like that intact, and it does not matter in the slightest that only a few people every year will go into it. That is precisely its value.[29]

All of these factors combined to create momentum in favor of the bill when John F. Kennedy arrived in the White House in 1961. Although Kennedy had little experience in wilderness himself, he recognized the gathering interest in the legislation and encouraged its supporters to press onward in Congress. In sharp contrast to Dwight D. Eisenhower, who had been silent on the legislation, Kennedy's backing of the bill was strong and meant additional support from the secretary of the interior, Stewart Udall, and the secretary of agriculture, Orville Freeman.[30] Kennedy's support also aided Democratic lawmakers from the West, including Idaho senator Frank Church, New Mexico senator Clinton

P. Anderson, and Montana representative Lee Metcalf, each of whom helped to shepherd the bill through the Congress. Republicans, notably Senator Thomas Kuchel of California and Congressman John Saylor of Pennsylvania, helped to champion the measure as well. In the political arena, wilderness enjoyed bipartisan support in the early 1960s to a far greater extent than it did during later decades.[31]

Momentum accelerated in favor of the bill when the Senate passed the legislation in 1961 and again the following year when President Kennedy accepted the report of the federal Outdoor Recreational Resources Review Commission. The fruit of five years of research and debate, the commission's report supported the establishment of wilderness areas to help meet the nation's soaring recreational demands. The report also provided political cover to western lawmakers who had been hearing a steady drumbeat of opposition to the bill from commodity producers. Wilderness areas, lawmakers also knew, were favored by sporting groups, rod and gun clubs, saddle and hiking organizations, and hunters and anglers. Big-game hunters from across the country had been flocking to the Rocky Mountain states, Pacific Northwest, and Alaska in ever-increasing numbers, and by the early 1960s they helped to generate increasing tourist income for residents living near the West's wilderness areas.[32]

Congress at last agreed to a wilderness bill in 1964. On September 3, President Lyndon Johnson signed the bill into law in the Rose Garden of the White House. The Wilderness Act declared, "[I]t is the policy of the Congress to secure for the American people of present and future generations the benefits of an enduring resource of wilderness." This landmark legislation established the national wilderness preservation system with 9.1 million acres permanently protected from roads, motorized vehicles, and equipment such as chainsaws. It set into motion a review process under which the National Forest Service and National Park Service had ten years to survey their potential wilderness areas and to offer recommendations to Congress. Primitive areas, though not formally included in the wilderness system, would be protected until Congress determined their ultimate status.

To be sure, wilderness supporters had been compelled to compromise. A key provision in the act permitted mineral leases and new mining claims to be filed on wilderness lands for twenty years until December 31, 1983. Grazing was permitted on wilderness lands where it already existed, and motorboats could be used in the Boundary Waters.[33] These compromises, coupled with the small amount of acreage initially established in the wilderness system, sparked criticisms of the Wilderness Act. Seeing it primarily as a zoning measure, critics charged that the law did little to instigate reforms of the forest service and its growing devotion to timber harvesting, which increasingly compromised wildlife habitat and protection of biological diversity. The act also did not compel Americans to curb their unquenchable appetite for natural resources or consumer goods. The Wilderness Act was not a reform measure and did not address patterns of consumption or social inequity that a later generation of environmentalists would see as crucial. More recently, some observers, informed by conservation biology, have viewed most wilderness areas as small islands of biological diversity too isolated from one another to enable endangered species to survive.[34]

Nevertheless, passage of the Wilderness Act proved a watershed in the nation's relationship to its wilderness. The act sparked years of activism by grassroots volunteers to gain passage of legislation setting aside more acreage in the wilderness system. In addition, the act became a touchstone for additional legislation, such as the Eastern Areas Wilderness Act of 1975, the Federal Land Policy and Management Act of 1976, and the Alaska National Interest Lands Conservation Act of 1980.[35] By 2006, more than 106 million acres had been designated as federally protected wilderness.

Another outcome was a nod to history. As wilderness preservation became more visible within the political and public arena, it seemed appropriate that the nation's historical relationship with wildlands should be chronicled. Historian Roderick Nash, whose father had served on the Outdoor Recreational Resources Review Commission, took up that challenge and published *Wilderness and the American Mind* in 1967. This work, a classic in American environmental history, helped to launch that new field of study while also inspiring generations of environmentalists, who took the Wilderness Act as a significant piece of evidence of their own achievements.

Aside from legislation and the political movement which arose to protect wilderness, the postwar generation of Americans who fought for the Wilderness Act captured the public's heightened interest in wildlands and wove this into the broader environmental movement. For them, preserving wilderness became a high moral cause tinged with religious commitment, a way to begin to set the human-nature relationship right. For some, it was a way of taking care of God's creation. In some respects, wilderness protection efforts provided a public lands counterpart to the civil rights movement which was unfolding at the same time.

This sense of high moral purpose and commitment to saving areas in which nature dominated was evident in the very language of the new law. "A wilderness," the act proclaims, "is hereby recognized as an area where the earth and its community of life are untrammeled by man, where man himself is a visitor who does not remain."[36] Wilderness preservation was an act of environmental responsibility, a sign of human commitment to other living things and their habitats. This notion of stewardship to guard the integrity of the natural world in all of its diversity and wonder became a core conviction of the emerging environmental movement in the 1960s, and it has remained at the center of the movement's principles in the decades since.

Notes

1. John Jakle, *The Tourist in 20th Century America* (Lincoln: University of Nebraska Press, 1985), 185–98; Wallace Stegner, *The Uneasy Chair: A Biography of Bernard DeVoto* (Garden City, N.Y.: Doubleday, 1974), 287–98.

2. See David Louter, *Windshield Wilderness: Cars, Roads, and Nature in Washington's National Parks* (Seattle: University of Washington Press, 2006).

3. Gregg Mitman, *Reel Nature: America's Romance with Wildlife on Film* (Cambridge, Mass.: Harvard University Press, 1999), 109–18; Margaret E. Murie, *Two in the Far North*, 2d ed. (Edmonds, Wash.: Alaska Northwest, 1978).

4. Edward Abbey, *Desert Solitaire: A Season in the Wilderness* (New York: Simon and Schuster, 1968), xiv, and *The Journey Home: Some Words in Defense of the American West* (New York: Penguin, 1977).

5. This notion of wilderness as a state of pure nature is part of what Michael P. Nelson and J. Baird Callicott mean by the "received wilderness idea" in their anthology, *The Great New Wilderness Debate: An Expansive Collection of Writing Defining Wilderness from John Muir to Gary Snyder* (Athens: University of Georgia Press, 1998).

6. Roderick Nash, *Wilderness and the American Mind*, 4th ed. (New Haven, Conn.: Yale University Press, 2001), 317–18; author's interview of Michael McCloskey, November 20, 1996, Sierra Club, Washington, D.C.; James Morton Turner, *The Promise of Wilderness: A History of American Environmental Politics* (Seattle: University of Washington Press, in press).

7. Adam M. Sowards, "William O. Douglas's Wilderness Politics: Public Protest and Committees of Correspondence," *Western Historical Quarterly* 37 (Spring 2006): 21–42.

8. E. A. Schilling to Regional Forester, Special Inspection Report, June 29, 1949, Wilderness Society Papers, Western History and Genealogy Department, Denver Public Library, Denver, Colorado (hereafter WS Papers).

9. Frank, "The Wilderness: A Major Water Resource," *Living Wilderness* 11 (June 1946): 5–16.

10. Bernard Frank and Anthony Netboy, *Water, Land, and People* (New York: Knopf, 1950).

11. Adam Rome, *The Bulldozer in the Countryside: Suburban Sprawl and the Rise of American Environmentalism* (Cambridge: Cambridge University Press, 2001); William Cronon, "The Trouble with Wilderness; or, Getting Back to the Wrong Nature," in his edited book *Uncommon Ground: Toward Reinventing Nature* (New York: Norton, 1995), 76–78.

12. "Voyageur's Return," *Nature Magazine* 41 (June–July 1948), 290. See also Grace Lee Nute, *The Voyageur's Highway: Minnesota's Border Lake Land* (St. Paul: Minnesota Historical Society, 1941).

13. Recent discussions of the sublime include Cronon, "The Trouble with Wilderness," 73–75; and Donald Worster, *A River Running West: The Life of John Wesley Powell* (New York: Oxford University Press, 2001), 306–8.

14. Howard Zahniser, "Cloud Peak's Primitive Area and People," *Living Wilderness* 12 (Autumn 1947): 15.

15. See Richard F. Fleck, "John Muir's Transcendental Imagery," in *John Muir: Life and Work*, ed. Sally M. Miller (Albuquerque: University of New Mexico Press, 1993), 136–51.

16. Zahniser's speech "The Need for Wilderness Areas" is available in Ed Zahniser, ed., *Where Wilderness Preservation Began: Adirondack Writings of Howard Zahniser* (Utica, N.Y.: North Country, 1992), 59–66; see also Mark Harvey, *Wilderness Forever: Howard Zahniser and the Path to the Wilderness Act* (Seattle: University of Washington Press, 2006), 90–92, 200–201.

17. R. Newell Searle, *Saving Quetico-Superior: A Land Set Apart* (St. Paul: Minnesota Historical Society Press, 1977), 165–86; David Backes, *A Wilderness Within: The Life of Sigurd F. Olson* (Minneapolis: University of Minnesota Press, 1997), 189–203; Mark Harvey, "Sound Politics: Wilderness, Recreation, and Motors in the Boundary Waters, 1945–1964," *Minnesota History* 58 (Fall 2002): 130–45.

18. See Mark Harvey, "The Changing Fortunes of the Big Dam Era in the American West," in *Fluid Arguments: Five Centuries of Western Water Conflict*, ed. Char Miller (Tucson: University of Arizona Press, 2001), 276–302.

19. The dam site itself was about two miles downstream from Echo Park in Whirlpool Canyon. Mark W. T. Harvey, *A Symbol of Wilderness: Echo Park and the American Conservation Movement* (Albuquerque: University of New Mexico Press, 1994; rpt., Seattle, University of Washington Press, 2000).

20. Wallace Stegner, ed., *This Is Dinosaur: Echo Park Country and Its Magic Rivers* (New York: Knopf, 1955; 2d ed., Boulder, Colo.: Roberts Rinehart, 1985); Harvey, *Symbol of Wilderness*, 235–59.

21. Harvey, *Symbol of Wilderness*, 190–205; Michael P. Cohen, *The History of the Sierra Club, 1892–1970* (San Francisco, Calif.: Sierra Club Books, 1988), 154–84, 357–65, 388; Byron Pearson, *Still the River Runs: Congress, the Sierra Club, and the Fight to Save Grand Canyon* (Tucson: University of Arizona Press, 2002).

22. Cohen, *History of the Sierra Club*, 291–99; *In Wildness Is the Preservation of the World* sold 70,000 hardcover and 250,000 in the first year of the paperback version; Jonathan Spaulding, *Ansel Adams and the American Landscape* (Berkeley and Los Angeles: University of California Press, 1995), 330.

23. Spaulding, *Adams and the American Landscape*; Finis Dunaway, *Natural Visions: The Power of Images in American Environmental Reform* (Chicago: University of Chicago Press, 2005).

24. Karl Onthank to Olaus Murie, Howard Zahniser, et al., Feb. 9, 1957, Box 1, Onthank Papers; "Decision of the Secretary of Agriculture Establishing the Three Sisters Wilderness Area," Box 45, Sigurd Olson Papers, Minnesota Historical Society, St. Paul, Minnesota; Kevin R. Marsh, " 'This Is Just the First Round': Designating Wilderness in

the Central Oregon Cascades, 1950–1964," *Oregon Historical Quarterly* 103 (Summer 2002): 210–33.

25. On the political battle, see Nash, *Wilderness and the American Mind*, 220–26; Harvey, *Wilderness Forever*, 186–244; Michael Frome, *Battle for the Wilderness*, rev. ed. (Salt Lake City: University of Utah Press, 1997); William L. Graf, *Wilderness Preservation and the Sagebrush Rebellions* (Savage, Md.: Rowman & Littlefield, 1990); Richard Allan Baker, *Conservation Politics: The Senate Career of Clinton P. Anderson* (Albuquerque: University of New Mexico Press, 1985); Craig W. Allin, *The Politics of Wilderness Preservation* (Westport, Conn.: Greenwood, 1982); and Doug Scott, *The Enduring Wilderness* (Golden, Colo.: Fulcrum, 2004).

26. Harvey, *Wilderness Forever*, 189, 196–98.

27. Quoted in Harvey, *Wilderness Forever*, 209.

28. Harvey, *Wilderness Forever*, 205–7, 223; Harvey Broome to Olaus Murie, July 30, 1961, Olaus Murie, 1961–64 folder, Harvey Broome Papers, McClung Historical Collection, Knoxville Public Library, Knoxville, Tennessee.

29. The "wilderness letter" appears in Page Stegner, ed., *Marking the Sparrow's Fall: Wallace Stegner's American West* (New York: Henry Holt, 1998), 111–17, quote on 116.

30. On Kennedy, see Steven C. Schulte, *Wayne Aspinall and the Shaping of the American West* (Boulder: University Press of Colorado, 2002), 115–51; see also Scott, *Enduring Wilderness*, 51.

31. Thomas G. Smith, *Green Republican: John Saylor and the Preservation of America's Wilderness* (Pittsburgh, Pa.: University of Pittsburgh Press, 2006).

32. In the Rocky Mountain and West Coast states, 3.4 million hunting and fishing licenses were issued in 1946; in 1958, 5.5 million; see Lawrence C. Merriam, "The Western States," *Proceedings Society of American Foresters Meeting, November 15–19, 1959, San Francisco, California* (Washington, D.C.: Society of American Foresters, 1960), 67–69.

33. See Michael McCloskey, "The Wilderness Act of 1964: Its Background and Meaning," *Oregon Law Review* 45 (June 1966): 288–321, which includes the text of the statute, Public Law 88–577, 88th Cong., September 3, 1964; Scott, *Enduring Wilderness*, 56–57.

34. For the role of wilderness areas within a new preservation strategy informed by conservation biology, see Dave Foreman, *Rewilding North America: A Vision for Conservation in the 21st Century* (Washington, D.C.: Island, 2004), 168–76.

35. For Alaska, see Edgar Wayburn with Allison Alsup, *Your Land and Mine: Evolution of a Conservationist* (San Francisco, Calif.: Sierra Club Books, 2004), 172–286.

36. Callicott and Nelson, *Great New Wilderness Debate*, 121; Howard Zahniser to Ned Graves, April 25, 1959, WS Papers; Douglas W. Scott, "'Untrammeled,' 'Wilderness Character,' and the Challenges of Wilderness Preservation," *Wild Earth* 11 (Fall–Winter 2001–2002), 72–79.

Twelve

Wilderness and Conservation Science

Michael Lewis

On April 21, 1948, an out-of-control trash fire spread in the sandhills of central Wisconsin. As fires go, it was a small one. No cities were threatened, and no large forests with accumulated years of dry timber stood nearby. It did threaten the farm on which it started, though, and the surrounding lots and forests. Neighbors, seeing the smoke in the air, went to help. One such neighbor was Aldo Leopold. Thirteen years earlier, Leopold had purchased a used-up, overfarmed, and abandoned piece of land in Sauk County. He and his family had spent their vacations driving up from Madison to restore the property to something approaching its pre-agricultural state, planting thousands of trees and nurturing the grasslands. On that April day, the fire was moving directly toward some of Leopold's rehabilitating meadows. He understood that some fires were good (and indeed, the prairie-oak savannah ecosystem of much of the upper Midwest depended upon them) and that there was a time and place for fires—but this was not one. The start of this fire was neither natural nor planned, and it was not burning in a prairie. This fire was fed by a landscape heavily modified by humans, and it would not be helpful in restoring ecosystem health. So he went to help put it out. The sixty-one-year-old Leopold suffered a heart attack in the midst of his exertions, and one of the twentieth century's most thoughtful conservationists died as we might expect he would have wanted—outside, working with his hands, confronting the power and wildness of nature even as he tried to control it in a landscape scarred by human misuse. He died, truly, of natural causes.

A year after Leopold's death, the book that has cemented his reputation as one of the great thinkers of the twentieth century, *A Sand County Almanac*, was published. The manuscript had been accepted for publication just before his death. This book illustrates some of the central tensions in how Americans have sought to understand and preserve wilderness—and particularly the emerging role of ecology in that process. Leopold held space in his conservation philosophy both for the preservation of vast wilderness areas (see chapter 10) and also for the restoration of completely degraded and all-too-human landscapes, as with his Sauk County property. And he pursued both simultaneously—the year 1935 saw his purchase of the Sauk County farm and his cofounding of the Wilderness Society. In the very first sentence of *Sand County Almanac*, written seven weeks before his death, he states, "[T]here are some who can live without wild things, and some who cannot. These essays are the delights and dilemmas of one who cannot." Significantly, Leopold found wild things not just in the wilderness of Gila National Forest, but in Sauk County as well, and he drew America's attention to both.[1]

Leopold's legacies are neither simple nor straightforward. Simultaneously a forester and a vocal advocate of wilderness, Leopold was the intellectual heir of both Pinchot and Muir, of both conservation and preservation, but he added to those traditions a more profound understanding of the developing science of ecology. While Paul Sutter in chapter 10 is right to point out that the new ecological sciences played only a small role in the interwar wilderness movement, by the 1940s findings from ecological studies done in the 1930s were transforming the way that Leopold (who was elected president of the Ecological Society of America in 1947) and other wilderness advocates understood the natural world. Leopold always maintained a sense of the human obligation to study nature, to attempt to understand and appreciate it in all of its complexity, and ultimately, to manage it when to do so would help to restore ecosystem health. Fittingly, he was Wisconsin's first professor of game management—he was the quintessential applied scientist.

The difficulty that postwar environmentalists have had in assigning Leopold a place in debates about the role of humans in wilderness (untouched wilderness, managed wilderness, or some mix?) mirrors a larger uncertainty about how best to understand science with regard to wilderness.[2] Many environmentalists have been suspicious of science, fearing that the scientific quest to learn the laws of nature is an attempt to master and control it—placing the human intellect above the workings of the natural world. For many environmentalists, this reeks of hubris, and they see the scientific revolution as a major cause of the environmental crises of modernity. That was certainly the opinion of activists in the 1960s who protested physics (the science of the atomic bomb), chemistry (the science of poison gas, toxic chemical compounds, pesticides, and herbicides), medicine and biology (associated with medical experiments and biological warfare), and the broad scientific underpinnings of industrialization and the military-industrial complex. And even ecologists have been known to destroy small bits of nature to serve their purposes, as when E. O. Wilson used cyanide gas to wipe out the animals on a tiny island as part of an experiment, or when Leopold cut down naturally sprouting trees that he did not want to grow on his Sauk County farm. The 1960s activists, though, believed that the science of ecology would be different—so much so that many claimed to be ushering in an "age of ecology" (by which they meant not the *study* of the environment, but the environment itself). And many ecologists came to agree, seeing their science as an attempt to understand, but never replace or control, nature. Following World War II, and particularly from the 1960s on, ecologists became environmental heroes, and ecological ideas became mainstays of environmental thought.

The science of ecology shaped how Americans understood, managed, and preserved wilderness in the second half of the twentieth century. Some rudimentary knowledge of the science of ecology mediated Americans' relationships with nature and wilderness. It supplied the examples they used to give meaning to what they saw and organized their expectations of how their actions would affect the earth. When Americans spoke of carrying capacity, or the balance of nature, or climax and succession, or the impact of extinctions, or invasive species, or predator-prey relationships, or edge effect, or islands of habitat, or biodiversity, or umbrella and keystone species, they used ideas that were expounded by ecologists. If anything, it is an understatement to claim that the science of ecology has shaped postwar U.S. environmentalism—the two at times seem to be completely intertwined.

The dominance of ecological science in environmentalism does not discount the importance of history and culture in shaping contemporary wilderness attitudes. The questions that ecologists ask and the ways that they attempt to answer them all are based upon their individual histories, their specific educational experiences, and their values. Further, both the funding and reception of scientific studies reflect the interests and values of the broader culture. These can be seen in the choices of animal subjects for ecological studies (more often grizzly bears than grey squirrels) and the landscapes that are selected for research (more frequently people-less wilderness areas than city parks).

Recognizing that culture-bearing humans conduct science does not diminish in any way the results of ecological studies conducted in the wilds, or on charismatic species—the information observed is not culturally determined. Eagles eat fish, carrion, and small mammals regardless of the cultural background of the scientist or how many times the birds are studied. But it does suggest that what we know about ecology is culturally focused in such a way as to emphasize certain species, landscapes, problems, and solutions. And in the context of the history of American wilderness, this means that U.S. ecologists and the science of ecology have often promoted a vision of wilderness more in keeping with Leopold's Gila Wilderness Area than his Sauk County property. In a feedback loop that would not look unfamiliar to scientists who study interactions between organisms and their environments, culture shapes science, which in turn produces knowledge that shapes culture.

This chapter looks at some of the ways in which conservation-oriented ecological sciences have reflected U.S. cultural attitudes about wilderness and have in turn contributed to the popular knowledge that Americans have applied in attempts to conserve wilderness areas. Throughout the twentieth century, the science of ecology and its many subfields have often suggested three things: (1) human influences upon nature are "disturbances" and undesirable; (2) wilderness is better preserved in large protected areas than in smaller ones; and (3) human civilizations are dependent upon nature (often glossed as "biodiversity") for their continued existence and health, yet human societies pose a grave threat to the continued existence of biodiversity. In this, ecology was congruent with existing American ideas about wilderness. The continued process of ecological research over the last fifty years—testing these assumptions against the reality of observed nature—has overturned some aspects of those culturally predicated convictions, while reinforcing others. In the last thirty years of the twentieth century, two distinct subfields of ecology emerged—conservation biology and restoration ecology—both drawing on different aspects of Leopold's work. Conservation biologists focused upon how to protect functioning ecosystems and wild species still in their natural habitats, and restoration biologists tried to restore shattered landscapes. By the start of the twenty-first century, a growing number of both groups of scientists came to agree with Leopold that wild areas must be preserved, but that removing human influences from the landscape was all but impossible. Human management and restoration based upon scientific principles—Leopold's "intelligent tinkering"—was essential. These scientists see wilderness as a blended space—both manipulated by humans in the interest of greater naturalness, yet still wild nature (and ultimately prone to "going wild" in directions unforeseen by managers). Leopold would certainly approve.

Disturbance and the Balance of Nature

Although the 1960s represented a coming of age of the popular awareness of the science of ecology, the science had been developing throughout the twentieth century. Ernst Haeckel coined the term "oecology" in 1867 in Germany to explain Darwin's ideas about the relationship between species and their environments. By the 1890s, biologists in both the United States and Europe began calling themselves "ecologists" to differentiate their work on communities of organisms from that of more species-specific colleagues in botany and zoology. There was a long tradition of scientists and naturalists who studied nature before ecology came into being, but as with the Bartrams in the eighteenth century, they usually did so by studying either plants or animals and usually with an eye to taxonomy, not interactions. The ecologists, while they still might tend to study primarily flora or fauna, differed in their explicit orientation toward interactions among communities of species. They began to see nature as a web of interactions, an ecological metaphor that has now become commonplace.[3]

The most influential of these early ecologists in the United States was Frederic Clements. Born in the prairies of Nebraska in 1874, Clements studied plant communities (he called them "associations") and developed the ideas of climax and succession. All climatic and geographical areas had a natural "climax" community of particular plant species. This climax community was assumed to be stable—the balance of nature. Of course, a fire, mudslide, or earthquake could cause a disturbance in the climax community. Barring the extinction of one or more species, the disturbed habitat would then undergo succession—an established, predictable pattern of regrowth. Thus, for example, a mature hardwood forest might burn in a forest fire. The deforested land would first come back as a meadow, followed by quick-growing shrubby species. Perhaps pines or other fast-growing softwoods would form a third stage, until eventually the hardwoods grew up between the pines, shading them out, and reasserting the climax community. Although many of his colleagues rejected his claim that these plant communities acted as a superorganism and evolved together, Clements's vision of static climax communities dominated ecology in the United States until his death in 1945.[4]

Clements's ideas have had even more staying power in environmentalist circles. Many environmentalists still affirm his central argument: nature, if undisturbed, would come to rest in a timeless and stable climax community (wilderness). Human manipulation of nature was, by definition, disturbance, and in fact, no climax community could include humans as they would always be interfering with natural processes. Further, if human disturbance were ended (barring human-induced extinction of species), nature would restore itself, via succession, to the climax community. The implications of these ideas for twentieth-century Americans concerned with the loss of wilderness in North America were obvious. All was not lost—if land were set aside and the human presence removed, nature would heal itself. Further, the presence of native peoples or rural settlers on the land need not invalidate its wilderness qualities—once the people were removed, time would heal all wounds. For the new discipline of ecology, not yet expert at analyzing past environments using contemporary tools, such as fossilized pollen analysis, charcoal identification, or the other methods of paleoecology, the idea of a climax community also gave its proponents confidence that if they went to a lightly impacted wilderness area, the environment they observed (in 1930, for instance) was not unlike what they might have seen elsewhere prior to the expansion of U.S. homesteaders. This motivated ecologists in the 1930s to pursue the preservation of areas with "natural conditions," as described in chapter 10.

Clearly, Clements's ecology provided one possible scientific justification for establishing people-less wilderness parks and for the philosophical preference for seeing nature as working best in the absence of humans. The continuing resonance of his theories has less to do with their scientific accuracy than with their correspondence to nonscientific wilderness values. Most environmentalists believe deeply in a balance of nature that is profoundly disturbed by the human presence.

Henry Gleason, also a rural midwesterner, was the ecologist most responsible for moving U.S. ecology past the idea of stable climax communities. Instead, Gleason suggested that "two patches of the same association [community] were never exactly alike. . . . Every variation of the environment, whether in space or in time . . . produces a corresponding variation in the structure of the vegetation."[5] There were no universal climax communities best suited to particular geographic or climatic areas. Instead, there were communities that reflected the historic ebb and flow of particular species, always expanding or contracting: a prairie only looks static if you are not looking closely, or not looking for long. Rather than a static climax, postwar ecologists increasingly saw constant change, succession with no end, and a focus upon the agency of individual species rather than community associations. Today, most ecologists reject any notion of a balance of nature as exemplified by an unchanging climax community, instead focusing upon collections of species constantly in struggle, with populations growing and falling, with unpredictable historical events (fires, floods, disease) constantly disrupting what equilibriums do develop.

Clearly, contemporary ecologists continue to value old-growth forests more than loblolly pine plantations, and most choose study sites with as little human disturbance as possible. Untangling the history of evolutionary pressures upon species—with the attendant questions of why some species thrive while others diminish, or why certain animal behaviors or plant characteristics emerge—is a central goal of many ecological studies. One could argue that human selection pressure (selecting for pine, for example, for its quick growth and soft wood that we can turn into paper) is "natural," insofar as human reproduction, growth, and consumerism emerged from nature. Most scientists (and environmentalists), though, choose to see human pressures as acting outside of "natural" evolution. Thus, to see evolution at work and to determine the "natural" workings of an ecosystem, an ecologist must work in an ecosystem with minimal human influence. Otherwise, an ecologist cannot determine if observed changes were induced by human pressures or the natural workings of the ecological community.

The strict statement that human-induced change is not natural can become confusing—for example, should eucalyptus trees be removed from California? They are native to Australia, but they have thrived in California. Traditional management practices in many state and national park systems call for managing protected areas with the goal of achieving the biological state of the year 1491—so eucalyptus would have to go. Yet this goal is often impractical. Removing these trees could potentially eliminate threatened bird species, already rare, which use the trees as habitat. Similarly, studies from different regions of the world have suggested that certain local ecosystems have evolved within the context of low-level human use—including the grazing of cattle—and that the elimination of this use would be destructive to other species that exist there.[6]

The Scientific Design of Wilderness Reserves

Perhaps the most important ecological theory to cross over into mainstream debates about preserving wilderness is the theory of island biogeography. In the 1960s, two ecologists, Robert MacArthur and E. O. Wilson, pooled their expertise to analyze the relationship between island size and species diversity. In 1967, they published *The Theory of Island Biogeography*, still one of the most cited works in the ecological literature.[7] From this study (based on research in the Caribbean), they derived the species-area curve, which shows a direct correlation between the area of an island and the diversity of plant and animal species that it can support. Wilson summed up the study in a simple phrase: "a reduction in habitat is inexorably followed by a loss of animal and plant species."[8]

The theory of island biogeography quickly passed from science to public advocacy. Other ecologists and environmentalists made the intuitive leap to suggest that what was true of oceanic islands was also true of terrestrial nature reserves, islands of "nature" in a sea of development or agriculture. Between 1969 and 1975, a growing number of people began to suggest that island biogeography demonstrated that nature reserves needed to be as large as possible in order to preserve the greatest number of species.

Another scientist, Jared Diamond, applied the (older) literature on edge effect to island biogeography to elaborate how species might be affected by island size.[9] The higher the ratio of edge to interior, the greater the impact of the edge on the species composition would be. The edge of a forest has different microclimate features than the center, with lower humidity, greater light, higher winds, and thus, slightly different vegetation patterns. Edges favored certain species (plants and animals both), and thus the addition of edges would shift the species composition of the area. If a bird required a dark forest for nesting, it would need to live on an island or reserve large enough so that the edge forest was not the whole island—and a circular reserve was therefore preferable to a pencil-shaped one.

It is important to note that island biogeography did not create the desire for large nature reserves, but it did provide scientific backing for this preference and thus reinforced it. The first ecologists who supported the application of island biogeography to conservation were already advocates of large reserves and used this theory to justify and solidify existing practice. In 1959, the United Nations Educational, Scientific and Cultural Organization (UNESCO) had asked the International Union for the Conservation of Nature, composed of scientists from around the world, for a list of the world's national parks and equivalent areas and what the criteria for these were. The resulting list, published in 1967, stated that "an area which is too small [is] not included." The scientists of the IUCN already believed that reserves should be large, but as the report came out in the same year as *The Theory of Island Biogeography*, it did not incorporate that theory. When the IUCN report was updated in 1973, it dramatically increased its size requirements for national parks and protected areas, using island biogeography as its justification.[10]

The theory of island biogeography complemented another rapidly developing theoretical approach used in justifying large reserves—minimum viable populations—which stemmed from the work of a number of scientists in the 1970s and 1980s.[11] Based upon genetics (and in-breeding effects), this theory suggested that in order for a species to survive for the long term (meaning thousands of years), there must be a minimum viable population. Several historical examples have seemed to support the assertion that many species become extinct once they fall below a certain population threshold. When populations reach a certain point, they are especially vulnerable to random events, such as a catastrophic fire or disease, or even a randomly skewed sex ratio for a few successive generations. There are also less obvious effects of low populations. Although it is commonly thought that passenger pigeons became extinct because hunters shot the very last ones out of the sky, it appears that they went extinct because they had behavioral triggers for breeding which were dependent upon large flocks of birds, and once hunters had reduced their population to a certain level, their breeding success plummeted.[12]

Although it seems intuitively obvious that below a certain population threshold species would go extinct, minimum viable population theories are largely untestable except through computer models and best guesses, because the time scale dealt with is usually a thousand years. As a leading advocate of minimum viable population models admitted in the 1980s, "Intuition, common sense, and the judicious use of available data are still state of the art." Thus, "minimum viable populations on the order of a few hundred to several thousand genetically effective individuals are within the range that satisf[ies] those scientists who have attempted to deal with real management situations."[13] This was apparently said with no irony, but there is a tremendous practical difference between having to save 300 or 5,000 tigers in order to preserve the species.

This theory fit with island biogeography in its emphasis on the need for large nature reserves. Scientists would calculate a minimum reserve size by multiplying the minimum viable population number by the amount of space each animal would need for its home range (often a large number, as many large carnivores, for instance, are highly territorial and cover a lot of ground looking for prey). For example, preserving the North American mountain lion with a population of 500 animals, with twenty-six square kilometers of territory each, would require a reserve of at least 13,000 square kilometers, about the size of Connecticut. This calculation is highly arbitrary, though, for both numbers are easily called into question. Scientists working on minimum viable populations admit that their best numbers are educated guesses. Similarly, how reliable are estimates of individual animals' home ranges? Some studies suggest that home ranges of carnivores are partially dependent upon the density of prey species. How can that yield a reliable range size for areas with different prey species numbers? Nonetheless, calculations of minimum viable populations, and the corresponding minimum viable reserve size, continue to be widely used in the scientific and popular literature about reserve design.

Ecologists were aware of the problems with minimum viable population figures, but without scientific rationales for conserving large reserves, two scientists feared that "pro-conservation individuals and groups, in and out of governments, hardly have a leg to stand on when competing for land and resources with powerful elements arguing for appealing, short-term or ill-conceived development activities."[14] The crisis of environmental destruction warranted judicious guesswork presented with scientific authority. The alternative seemed to be to give up the fight to those who would develop every natural area and clear-cut every forest.

The implications of island biogeography and minimum viable population estimates should be obvious with regard to the conservation of wilderness areas. Aldo Leopold had suggested that wilderness areas should be able to support a two-week pack trip (with a mule). Island biogeography and minimum viable populations could give a much more scientific, even mathematical, definition. Island biogeography could be used to maintain all of the species in a wilderness area (species diversity), or minimum viable population estimates could be used to maintain a particular species. Many wilderness advocates (and scientists) quickly realized that cataloging the complete biodiversity of any ecosystem is daunting, especially when it is large. They therefore turned to the notion of umbrella and keystone species. Conservation based upon umbrella species argued for reserves large enough for the most broadly roaming species in that ecosystem (usually the large carnivores; they did not include migratory birds), and then the whole ecosystem could be maintained. The keystone species idea argued that one particular species is particularly important in maintaining the diversity and health of the complete ecosystem—usually, ecologists have focused upon top carnivores as keystone species. Conservation based upon umbrella and keystone species justified focusing upon conserving only one species and assuming that all others in the ecosystem would be protected as well, and it explicitly used minimum viable population estimates as its basis. This species-based approach also fit well with environmental advocacy. It is easier to mobilize support for a charismatic large mammal (such as a panda bear or a tiger) than for the whole ecosystem in which they live, or for less charismatic but potentially more endangered species. Even as environmentalists increasingly speak of biodiversity, it is easier to design reserves based around large mammals.

There have been ecologists who challenged the assertion that large reserves would necessarily contain a greater number of species than smaller ones.[15] They argued that, if you had a hundred square kilometers to devote to a national park, it would be better to have ten parks of ten square kilometers than one big park. Doing so would allow for coverage of more "microhabitats," meaning a greater diversity of land types and ecosystems, and would avoid putting all of the biological eggs in one nature preserve basket (in the event of disease or fire). If the parks were sited close together or had linking corridors, the animals could still move around and recolonize the other small parks, which might have suffered local extinctions. Although small parks might not accommodate large mammals, several small parks would almost certainly guarantee that a greater number of species of insects and plants would be preserved. Given that even supporters of large nature reserves sometimes worried that "species extinction of the large vertebrates seems inevitable,"[16] why should the most effort be put into preserving those very species with large reserves? This dispute became known as the SLOSS (single large or several small) debate, and it raged through ecological journals from the 1970s to the 1990s.

Is it sufficient to simply preserve maximum biodiversity, or must pristine people-less naturally functioning ecosystems—wilderness, traditionally defined—be the goal? Although small reserves often maintain a high biodiversity, they have to be managed in an unnatural fashion (for instance, culling certain animals when the population becomes too high, as with deer throughout most of the eastern United States). Proponents of large reserves have seized upon this aggressive management as a defect. Many ecologists are strongly committed to the idea that naturally functioning ecosystems are worth saving in and of themselves and that any human involvement in any ecosystem, by definition, disturbs the normal or "natural" workings of that system. Arguing for large reserves is arguing for ecosystems of a sufficient size so that human management is not needed.

Implicit in the several-small position is the idea of small nature parks scattered in a matrix of human use. This is anathema to the scientist who sees any human contact with the natural world as harmful or who hopes to see in nature the working of evolution without human influence and to the wilderness advocate who sees wilderness as operating outside of human control. By the 1990s, most scientists admitted that SLOSS was an unsolvable debate, because proponents began from different basic value positions. Large and small reserves are both appropriate in different contexts and for different goals. Any ecologist would admit that a huge park does not have any inherent *ecological* disadvantages, but such a park is often not politically or socially preferable, let alone feasible. Though most supporters of large parks agree that more total species would be saved by carefully placed small parks in a variety of habitats, they point out that such parks cannot sustain larger animals nor many of the nongeneralist smaller species. And such small reserves do not save wilderness, in the traditional American view.

Extinctions and Conservation-Oriented Science

Ecology did not just contribute ideas to U.S. environmentalism—it also supplied some of the postwar period's leading environmental advocates, though initially only the rare few scientists were willing to leave their laboratories or fieldwork to directly address the public. Rachel Carson, a retired U.S. Fish and Wildlife Service employee, is often credited with starting the mass environmental movement (as distinguished from the earlier wilderness movement) with her landmark 1962 book, *Silent Spring*. Carson brought together a wide array of published and unpublished ecological studies on the effects of chemical pesticides on specific species and the environment. Her central thesis was explicitly ecological: human pesticide use runs the risk of destroying the very environment that makes human life possible. All life is interconnected, she warned, and any human manipulation of the ecosystem is fraught with danger. A second key environmentalist tract of the 1960s, ecologist Paul Ehrlich's bestselling *The Population Bomb*, warned of the dangers to the earth of human overpopulation. Though his own research was on insect populations, he extrapolated his findings to the ecological ramifications of the human species overrunning its habitat. As Americans became convinced that they faced a series of environmental catastrophes—from the destruction of fragile wilderness areas, to the extinction of charismatic species, to widespread pollution and attendant human health risks—they needed credible information upon which to base solutions. As Carson and Ehrlich demonstrated, there was a voracious popular appetite for general scientific tracts explaining environmental problems.

Carson and Ehrlich (and a few other colleagues) were the exception in the 1960s. Far more typical were scientists who eschewed environmental advocacy, believing that to become an "environmentalist" was to abandon scientific objectivity.[17] No scientist better illustrated this than E. O. Wilson. As the SLOSS wars (which his theories had helped to start) raged around him in the latter half of the 1970s, Wilson ignored them. As he admits in his autobiography, he looked down at his ants and took "some relief from the knowledge that non-academic organizations were already active in the conservation of biological diversity."[18] He relied upon others to popularize the relevant ecological ideas, as they had with island biogeography.

In 1979, the British ecologist Norman Myers ushered in a new age of scientific activism with the publication of *The Sinking Ark: A New Look at the Problem of Disappearing Species.* Myers took a familiar theme for ecologists—extinction—and made it into a crisis. Wilson's theory of island biogeography was based on extinction rates, but it was a theory. Myers "published the first estimates of the rate of destruction of tropical rain forests."[19] Globally, 1 percent of the rainforest was disappearing every year, Myers claimed, and he tied this loss of habitat to the extinction of species. He claimed that one million species would be lost between 1975 and 2000.[20]

This changed Wilson's nonchalance; as he remembers, "this piece of bad news immediately caught the attention of conservationists around the world," including himself.[21] Ecologists in the United States had a long-standing appreciation for tropical forests and tropical ecology. When Myers wrote of the loss of the world's rainforests, he was writing of the loss of Wilson's and many other ecologists' favorite biota and field site. When he wrote of the loss of insect diversity, he wrote of Wilson's study species. Wilson and many of his colleagues became environmental advocates, and they did so because of Myers's description of a habitat loss based extinction crisis. They joined conservation boards. They went to the media with wildly different estimates of the global rate of extinction; they popularized the plight of particular species. They published popular books, and articles by and about them appeared in popular magazines. They became some of the most recognizable environmentalists in the United States, from Jane Goodall to Wilson himself.

There were historical precedents for U.S. ecologists taking on the mission of preserving the world's remaining wild areas, particularly the habitats of endangered species and biodiversity-rich tropical forests. In 1925, Albert National Park, designed to protect gorillas in the Belgian Congo, had been established by King Albert at the urging of U.S. scientist Carl Akeley, who died while conducting research there. By 1927, the Smithsonian Institution alone had sponsored more than 1,500 scientific expeditions across the globe.[22] Throughout the twentieth century, field ecology was wrapped in an aura of discovery and adventure, and it allowed American scientists to experience a (global) frontier in the world's wildernesses. Their descriptions of why they went into the field could have been written by John Muir or, for some of them, Theodore Roosevelt. As the globe-trotting E. O. Wilson wrote, "[N]ature [was] a sanctuary and a realm of boundless adventure; the fewer people in it, the better. Wilderness became a dream of privacy, safety, control, and freedom. Its essence is captured for me by its Latin name, solitude."[23] Just as some earlier frontier-loving Americans went to the wilds to challenge themselves with the strenuous life, as described in chapter 9, biologist Alan Rabinowitz, a global leader in jaguar and tiger conservation, was profiled in *National Geographic Adventure* as a shy, stuttering loner who went to the woods to study black bears and there "found his way, both as a scientist and as a man." The article is full of details about Rabinowitz's conservation successes, his science, his ongoing battle with cancer, but also his manliness—from his weight lifting, to his skill at Thai sword fighting, to the time that he arrived at a Burmese village and to establish credibility "beat the village's strongest man in an arm-wrestling competition."[24] Wilderness conservation by ecologists overseas was not so different than wilderness conservation in the U.S. West a few decades earlier, but now scientific rationales were overlaid on the earlier cultural beliefs about the importance of nature as a sanctuary and a testing ground.

By 1986, a group of biologists who were actively working toward conservation goals decided to found the Society for Conservation Biology (SCB), formally recognizing conservation-oriented ecology (this was also the year in which the word "biodiversity" was invented by Wilson and another ecologist as a self-conscious marketing tool). Michael Soulé, a leading proponent of minimum viable population theories and a founder of the SCB, described conservation biology as "the application of science to conservation problems, [addressing] the biology of species, communities, and ecosystems that are perturbed, either directly or indirectly, by human activities or other agents. Its goal is to provide principles and tools for preserving biological diversity."[25] Put more simply, conservation biology was "a friendly, mission-oriented science that justifies the necessity for large areas of inter-connected wilderness."[26] This field was explicitly oriented around island biogeography and minimum viable population estimates, in order to slow down the extinction crisis.

Although the subdiscipline was formalized in 1986 and based upon theories that had emerged in the 1960s, people had practiced conservation biology much earlier. Although the SCB gave ecologists a new platform from which to claim expertise in preserving wildlands and wild species, the underlying goals of species and ecosystem conservation were scarcely different than those being advocated by some ecologists and wildlife biologists decades earlier. The naming of conservation biology, and the association created in its name, was not revolutionary, but rather the final political step in moving ecological science into the service of conservation goals.

At nearly the same moment as conservation biology was being formalized as an academic subfield in the 1980s, restoration ecology was emerging from a different scientific tradition within ecology. The Society for Ecological Restoration was founded in 1987, and the leading journal *Restoration Ecology* was founded in 1993. As with conservation biology, there were long antecedents for restoration ecology, but while conservation biology was often based in wildlife biology, restoration ecology was often derived from applied plant studies (including the long history of scientific attempts to restore forests and grasslands). Restoration ecologists work at several different levels: some focus on restoring specific species, others on communities, still others on entire ecosystems or landscapes. While conservation biology experienced explosive growth in the 1980s, restoration ecology did not see a similar growth in interest and practice until the mid-1990s.[27]

Restoration ecology and conservation biology are sometimes presented as rival ways of understanding the human role in nature. Restoration ecology presupposes that humans can and must manage nature; most conservation biologists have argued for attempting to let nature manage itself through the removal of human pressures and influences. Conservation biologists have historically focused upon saving ecosystems that are not yet degraded and species that are not yet extinct; restoration ecologists focus upon degraded and abandoned landscapes and remnant species, with the goal of restoring them. One scientist claimed that restoration ecologists were "optimistic," conservation biologists "pessimistic," with regard to viewing the status of the natural world.[28] Thoughtful scientists and environmentalists, though, have seen the need for both sciences. A restoration ecologist comparing the two subfields argued that the world is currently passing through a devastating bottleneck in which many species are going extinct and many ecosystems are being destroyed. Conservation biology focuses upon widening that bottleneck as much as possible, while restoration ecology proposes shortening the bottleneck. Both are essential.[29] But is a restored landscape wilderness? Or is wilderness lost as soon as human scientists direct its growth, its restoration? By the twenty-first century, even environmentalists such as Dave Foreman, founder of Earth First! and the Wildlands Project (discussed in more detail in chapter 14), have moved to accept that humans will have to manage wild areas. In 2004, Foreman published *Rewilding North America: A Vision for Conservation in North America*. This most ardent of wilderness advocates accepts that the extinction crisis, combined with other human impacts to the biosphere, has grown so severe that simply setting aside land is insufficient for preserving wilderness. He became the director of the Rewilding Institute, a think tank that includes conservation biologist Michael Soulé among its key members and that dreams of the large-scale rehabilitation of North American wilderness in massive interconnected refuges with human reintroductions of key species (particularly large mammals).

The Aldo Leopold Wilderness Research Institute

In Montana, a unique consortium of scientists and social scientists supported by various government agencies and the University of Montana study these same questions about wilderness management at the appropriately named Aldo Leopold Wilderness Research Institute. At the heart of their research is a dilemma that reflects the history of wilderness preservation in the United States. Wilderness, according to the Wilderness Act of 1964, should be "untrammeled by man." Later in that same bill, the Congress declared that wilderness is to be "managed so as to preserve its natural conditions." As scientist Peter Landres and his colleagues at the institute ask, "How do you manage wilderness so that it is both wild and natural?"[30] Yellowstone National Park provides several examples of how this debate matters. Elk populations are high in Yellowstone, and they eat young aspen trees. To leave Yellowstone wild (and not manage the elk population) would result in a potential loss of aspen trees. One reason that the elk populations are so high is the elimination of wolves and human hunters (native tribes used to hunt there, as described in chapter 6). Clearly, few people want to reintroduce hunting in Yellowstone, though to do so might be "natural," in the sense that low-level human hunting occurred in Yellowstone for centuries. The reintroduction of wolves is more palatable to environmentalists—though that action is a clear case of human management of the ecosystem. As Landres states, "Should the wildness of present-day wilderness be compromised to restore naturalness? In other words, should an undesirable means, such as manipulation of wilderness, be used to achieve a desirable end, such as restoration of natural conditions in wilderness?"[31] In most wilderness areas, this human management will have to be ongoing. The decision to not manage wilderness areas will result in far different landscapes emerging, often dominated by invasive, opportunistic, or exotic species (weed species, as they are sometimes called) that were not even present two hundred years ago.

In order to make management decisions, scientists and managers must decide what wilderness is and what they want it to be. These questions are answered not just by science, but also by cultural values. For much of the twentieth century, it was possible for scientists to act as if the definition of wilderness, or natural, was self-apparent—it meant nonhuman. But by the late twentieth century, ecologists began to consider some of the contradictions inherent in the preservation and restoration of wilderness, and they realized how degraded and unstable many preserved ecosystems already were. "Natural management," in which there is no human intervention in the workings of nature, was adopted as the ideal for U.S. national parks in 1969 and applied in our largest nature reserves, such as Yellowstone. By the 1990s, natural management was recognized by most ecologists and park managers to be insufficient.[32] As the twenty-first century begins, it is difficult to imagine a wilderness in which wildness and human thought or action will not be blended. The choice is not between pure or impure wilderness, but between a wilderness in which we attempt to choose our influences and one in which our influences are unplanned and uncontrolled. And as science has shown us, even when we attempt to plan our influences, we cannot absolutely know what will happen. Nature still bats last.

Notes

1. Susan Flader, *Thinking Like a Mountain: Aldo Leopold and the Evolution of an Ecological Attitude toward Deer, Wolves, and Forests* (Columbia: University of Missouri Press, 1974); and Curt Meine, *Aldo Leopold: His Life and Work* (Madison: University of Wisconsin Press, 1988).

2. See J. Baird Callicott and Michael Nelson, eds., *The Great New Wilderness Debate* (Athens: University of Georgia Press, 1998), for numerous battles over Leopold's legacies.

3. See Donald Worster, *Nature's Economy: A History of Ecological Ideas*, 2d ed. (New York: Cambridge University Press, 1994), for the history of ecology.

4. For Clements and Gleason (discussed in the next paragraph), see Michael Barbour, "Ecological Fragmentation in the Fifties," in *Uncommon Ground: Toward Reinventing Nature*, ed. William Cronon (New York: Norton, 1995), 233–55.

5. Henry Gleason, "Autobiographical Letter," *Bulletin of the Ecological Society of America* 34 (1953): 40–42, as quoted in Barbour, "Ecological Fragmentation," 237.

6. These concepts are explored in more detail in Michael Lewis, *Inventing Global Ecology* (Athens: Ohio University Press, 2004).

7. R. H. MacArthur and E. O. Wilson, *The Theory of Island Biogeography* (Princeton, N.J.: Princeton University Press, 1967).

8. E. O. Wilson, *Naturalist* (Washington, D.C.: Island, 1994), 355. See Lewis, *Inventing Global Ecology*; and Stephen Budiansky, *Nature's Keepers: The New Science of Nature Management* (New York: Free Press, 1995), for more detailed analyses of these theories.

9. See Jared Diamond, "The Island Dilemma: Lessons of Modern Biogeographic Studies for the Design of Natural Reserves," *Biological Conservation* 7 (1975): 129–46.

10. Richard Forster, *Planning for Man and Nature in National Parks: Reconciling Perpetuation and Use* (Morges, Switzerland: IUCN, 1973), 26.

11. One key study is R. J. Berry, "Conservation Aspects of the Genetical Constitution of Populations," in *The Scientific Management of Animal and Plant Communities for Conservation*, ed. E. D. Duffey and A. S. Watt (Oxford: Blackwell, 1971), 177–206. More recent work is by Michael Soulé and B. A. Wilcox, *Conservation Biology: An Evolutionary-Ecological Perspective* (Sunderland, Mass.: Sinauer, 1980).

12. Paul Ehrlich and Ann Ehrlich, *Extinction: The Causes and Consequences of the Disappearance of Species* (New York: Random House, 1981).

13. Michael Soulé and Daniel Simberloff, "What Do Genetics and Ecology Tell Us about the Design of Nature Reserves," *Biological Conservation* 35 (1986): 32.

14. Soulé and Simberloff, "Genetics and Ecology," 35.

15. Daniel Simberloff and L. G. Abele, "Island Biogeography and Conservation Strategy and Limitations," *Science* 193 (1976): 1032.

16. Francesco Di Castri, F. W. G. Baker, and Malcolm Hadley, eds., *Ecology in Practice*, vol. 1, *Ecosystem Management* (Paris: UNESCO, 1984), 414.

17. Sara Tjossem, "Preservation of Nature and Academic Respectability: Tensions in the Ecological Society of America, 1915–1979" (Ph.D. diss., Cornell University, 1993); David Takacs, *The Idea of Biodiversity* (Baltimore, Md.: Johns Hopkins University Press, 1996), 4.

18. Wilson, *Naturalist*, 356.

19. Wilson, *Naturalist*, 357.

20. Norman Myers, *The Sinking Ark: A New Look at the Problem of Disappearing Species* (New York: Pergamon, 1979).

21. Wilson, *Naturalist*, 357.

22. Lewis, *Inventing Global Ecology*, 71.

23. Wilson, *Naturalist*, 56.

24. Michael Shnayerson, "The Fight of His Life," *National Geographic Adventure* 7, no. 10 (December 2005–January 2006): 57, 99.

25. Michael Soulé, "What Is Conservation Biology?" *BioScience* 35, no. 11 (December 1985): 727.

26. Michael Soulé, introduction to Edward Grumbine, *Ghost Bears: Exploring the Biodiversity Crisis* (Washington, D.C.: Island, 1992), xi.

27. Joan G. Ehrenfeld, "Defining the Limits of Restoration: The Need for Realistic Goals," *Restoration Ecology* 8, no. 1 (March 2000): 2–9; and Truman P. Young, "Restoration Ecology and Conservation Biology," *Biological Conservation* 92 (2000): 73–83.

28. Young, "Restoration Ecology and Conservation Biology," 79.

29. Young, "Restoration Ecology and Conservation Biology," 77.

30. Peter B. Landres et al., "Naturalness and Wildness: The Dilemma and Irony of Managing Wilderness," in *Wilderness Science in a Time of Change Conference*, vol. 5, *Wilderness Ecosystems, Threats, and Management*, comp. David N. Cold, Stephen F. McCool, William T. Borrie, and Jennifer O'Loughlin (Ogden, Utah: U.S. Department of Agriculture, Forest Service, Rocky Mountain Research Station, 2000), 377–81.

31. Landres et al., "Naturalness and Wildness," 378.

32. Lewis, *Inventing Global Ecology*, 212–16.

Thirteen

Creating Wild Places from Domesticated Landscapes
The Internationalization of the American Wilderness Concept

Christopher Conte

There is a complex history and geography behind the ways that wilderness ideas have permeated particular places across the globe. My own experience with the global forces of wilderness protection comes from the Amani Nature Reserve (ANR) in Tanzania's East Usambara Mountains. In 1997, the Tanzanian government formally established the ANR in order to protect a patchwork of biologically rich forests that had eluded the development of the timber and tea industries, which began to exploit the area in the 1950s. With the rhetorical inclusiveness now characteristic of the international conservation lobby, the law designated the forest preserves for the material, scientific, and ecological benefit of all "stakeholders."[1] The government called for local participation in forest conservation, while arguing that the ANR's pristine wilderness attributes should lead to its international recognition as a biosphere reserve or a world heritage site, a status which would further impinge upon indigenous power to determine how the forest might be used.

I visited the area in 1998 as part of a research project investigating a 1940s famine that had pushed a number of refugees out of more heavily populated neighboring districts and into the less-stressed East Usambara Mountains. Despite my best efforts to focus the conversations on famine history, almost all of the interview sessions turned to the subject of the new ANR, which seemed to many an immediate and palpable threat to their livelihoods. Inside the reserve, farmers continued to cultivate cash crops and to collect firewood while children captured live reptiles for sale in Europe. Such activities seemed both to confirm the strong ties between the forest and its bordering communities and to contravene the preservationist spirit and letter of the new government decree. In this case, the obvious need for people to supplement their farming livelihood with forest commodities had complicated conservation efforts, and conflicts over access to the forest remained unresolved in 1998.

In 2004, somebody discovered a gold deposit in a stream bed just outside the ANR boundary and tens of thousand of miners, most of them unlicensed, flocked to the East Usambara Mountains. Once prospectors had laid claim to the deposits outside the ANR boundary, newcomers began to enter the reserve forests illegally and, in the throes of gold fever, they destroyed several stream sources.[2] Now that the gold is gone, so are the miners, and the social situation has returned to "normal." In their wake, however, they left a great deal of environmental damage.

The forests inside and outside the ANR, as well as the people who know them and use them, are caught in a pincer between the Tanzanian government and wilderness conservation advocates, on the one side, and the economic realities of Tanzania's poverty, on the other. Under these circumstances, international, regional, and local forces tear at the forest ecosystem. The ANR designation represents the culmination of a historical process wherein foreign notions of wilderness have been recently layered onto a place where people have been farming for more than a thousand years. The ANR legislation restricts rights of use and occupation that have developed locally over the very long term, while it ignores the realities of the more recent intrusion of the global economy. The tense situation at Amani demonstrates that wilderness legislation based on external models alone will not protect these clearly valuable forests. Nor will laying the blame for forest degradation exclusively on local farmers. My historical analysis of environmental change in the region implicates international logging interests, rather than rural land-use practices, as the major purveyors of forest degradation.[3]

This chapter illustrates some of the impacts of the nineteenth-century American wilderness idea in such international contexts, as well as some of the resonances between U.S. wilderness history and European colonial history. In places as diverse as Tanzania, Indonesia, New Zealand, Tasmania, and Brazil, among many others, conservation policy and practice exhibit strong ties to an American legacy of wilderness designation and protection. However, in these international contexts, American wilderness ideology, itself a historical product of foreign and homegrown conceptualizations of nature, has combined with the biological sciences and the legacy of colonial and postcolonial authoritarian rule to produce landscapes divided by conflict. The historical perspective offered below demonstrates that the application to these places of wilderness conceptualizations based on notions of unspoiled nature forms part of the history of continuing Western imperialism. This chapter argues, furthermore, that the environmental histories of wildernesses show humanity's formative role in these landscapes.

Science forms one of the key links between imperialism and wilderness conceptualizations. In part, this chapter examines the foundational role of nineteenth-century natural history's holistic vision in the development of ecological science and the idea of wilderness. The natural historians, with their philosophical grounding in romanticism, helped to inspire the wilderness ideal that later shaped American conservation science, as seen in the previous chapter. Internationally, writers like Alexander von Humboldt, Charles Darwin, and Alfred Russel Wallace helped to reify the idea that lands outside Europe and North America were places of wild natural beauty and wonder. While their writings demonstrate a decided ambivalence toward the social, economic, and cultural lives of the human communities they observed living in the tropics, indigenous peoples nonetheless occupied a space in their imagery and their thinking. The increasing sophistication of the biological sciences, though, led in the late nineteenth century to the demise of the holistic natural history approach. What remained was a systematic classificatory system; a concern with the interactions of the physical, chemical, and biological worlds; and an evolutionary perspective on nature in which humanity played little, if any, part. These early transformative processes have influenced the modern scientific study of ecosystems, as well as American practices and philosophies of wilderness protection. Thus, in the colonial world, the joint ethics of biological science and wilderness protection provided a rationale to create parks and game reserves that, when combined with the authoritarian legacy of colonialism, resulted in places controlled by bureaucracies rather than local people. To many of these local people, the American wilderness idea has been less a way of seeing nature than a government practice, a mode of land management.

This chapter's five sections aim to unravel the processes that resulted in authoritarian wilderness protection in the international sphere. The first section examines the ways that societies and cultures view landscapes, arguing that outside observers often see wildness in the same places where indigenous peoples see domestication. The second section outlines the role of nineteenth- and early twentieth-century naturalists in defining wildness in tropical landscapes. The third section describes how colonial states imposed their concept of wilderness value on indigenous spaces, often with disastrous results for the peoples whom they evicted from the newly formed parks. By adopting a historical perspective, the fourth section demonstrates the human past in wilderness landscapes. Finally, I argue that wilderness conservation has reached an impasse, which could be overcome through new scientific and conceptual ideas springing from local perspectives.

Conceptualizing Wilderness: The Landscape Perspective

Any discussion of the internationalization of the wilderness concept begs the question of just where it came from. The most ubiquitous model of wilderness protection arose in the nineteenth-century United States, and it continues to set the global example for the strict preservation of nature through the American national parks system. Yellowstone National Park, founded in 1872, was originally created not for the preservation of wildness, but for the exploitation of curiosities such as geysers, hot springs, and waterfalls.[4] Shoshone Indians continued to live in the park while other Native American groups hunted and fished there seasonally under the auspices of park managers and a military contingent. However, in a move foreshadowing similar trends across the globe, in the interest of preserving the park's wilderness, all indigenous peoples were eventually evicted from Yellowstone—as discussed by Angela Miller and Benjamin Johnson in chapters 6 and 7. And, as also shown in preceding chapters, by the late nineteenth century, most Americans associated national parks with a specific vision of people-free wildernesses that prohibited settlement and subsistence and commercial use of natural resources. This vision of national parks was best expressed in John Muir's glorification of wild nature at Yosemite in California's Sierra Nevada, and parks came to be recreational, aesthetic, spiritual, and biological reserves.

This type of wilderness national park caught the industrialized world's imagination in the nineteenth century. Whether by King Albert touring the parks of the West and then declaring a similar park in the Belgian Congo or Tsuyoshi Tamura visiting Yosemite and then working to create Japanese national parks, among many such examples, the fundamental philosophy and practice of American wilderness protection via national parks has been applied globally with remarkable persistence. Since the 1930s, U.S. and international environmental organizations have promoted this with direct economic and political support.[5] In 1962, the United States hosted the first of a series of World Conferences on National Parks, explicitly designed to spread the national park model globally. Well attended by government officials and activists from throughout the world, these conferences glorified the U.S. model, and were successful.[6] Unfortunately for indigenous peoples, the preservationist viewpoints that evolved in the

American West required the removal of Native Americans from the parks, as well as an intellectual leap that erased their historical influence on the parks' ecosystems. Thus, wilderness space, as defined by outsiders, became space devoid of humanity's imprint. Not surprisingly, many Europeans attempted to apply the U.S. national park model not at home, but in their colonies.

Gary Nabhan, whose work examines the dynamic interrelationships of indigenous culture and nature in the North American Sonoran Desert, has developed a useful way to contrast the ways that people view landscapes. His "cultural parallax" differentiates the viewpoints of those who live in and shape the habitats of their home range and outsiders who view the same spaces as "landscapes."[7] Nabhan's work demonstrates the intricacies of human relationships that develop when people come to know a place's plants, animals, and landforms, as well as its productive possibilities, through generations of actively managing their homes.[8]

From the insider viewpoint that Nabhan provides, readers begin to see how, over many generations, a group of indigenous Americans domesticated a desert in what appears at first glance to be an isolated and uninhabitable corner of the U.S.–Mexico borderlands. This essentially historical process eludes outsiders who, without knowledge of the past, cannot see on the land the many layers of cultural and natural complexity. In such places, visitors see wildness rather than domesticity, emptiness rather than habitation, while Nabhan's point of view animates this desert environment with cultural life and with power. Despite the cultural and natural beauty that Nabhan describes, the Papago Indians' rich store of knowledge teeters on the brink of extinction under the influence of the homogenizing influences of Mexican and American cultures. Nineteenth- and twentieth-century European colonialism led to a similar process whereby a group of outsiders developed a landscape perspective that reflected the interests of settlers, research scientists, and the colonial bureaucracy. Whether their motives involved preservation or exploitation, the complexities of indigenous influence on the land often remained invisible to them.[9] Thus, in the colonial context, places that were layered with multiple meanings for local peoples became for colonists empty spaces—wildernesses—that required a new geographical and botanical classification.

A landscape is a product of lived experience and is therefore an essentially historical entity.[10] Paradoxically, in colonial contexts, wilderness designations such as wildlife parks, game reserves, forests, and national parks removed from these landscapes their meaningful human history. The colonial mindset ascribed value to wilderness reserves for their *natural* history and therefore for their lack of a cultural past. Colonialism thus defined and delineated "wilderness" spaces where before there had been native landscapes. It remained for colonial scientists,

European hunters and settlers, and wildlife managers to provide these spaces with complexity and history.

Seeking Wildernesses

The tradition of natural history helped both to identify resources for imperialist exploitation and to shape the eventual scientific justification for their conservation. Eighteenth- and nineteenth-century naturalists were generalists, concerned not only with observing and describing nature, but also with the ethnographic classification of the peoples whom they observed within it.[11] Their holistic vision blended culture and nature into an intellectual world view that isolated indigenous societies, economies, and environments in time and space. In this way, European and American naturalists described people from places as diverse as Alaska, the Colorado Plateau, Amazonia, Papua New Guinea, the South Pacific islands, and Africa as living in a natural state that the incursion of Western civilization would eventually destroy. Whether they described their subjects as benighted or blessed, the idea that prior to European or American colonization indigenous peoples had lived in homeostasis with nature, and therefore without history, proved remarkably persistent (as with the "noble savages" of chapter 2).

Perhaps the most important natural historian was the German Alexander von Humboldt (1769–1859), who inspired a succession of American and European naturalists, including John Muir, who aspired to be a "new Humboldt" and who carried writings by Humboldt with him on some of his early wanderings.[12] Humboldt and his disciples collectively developed a historical vision of nature that would greatly influence ecological science (and obviously, through Muir and others, the American wilderness idea). Humboldt's overriding passion was to measure what he believed to be nature's dynamic equilibrium, which he argued derived from an infinite complexity, rather than a benevolent controlling hand.[13] Despite the imprint of Humboldt's conceptualization of equilibrium on natural history, no consensus emerged regarding humanity's role in nature. Henry Walter Bates, who spent eleven years (1848–1859) traveling, observing, and celebrating Amazonian nature, for example, contrasted the wealth of tropical nature with the poverty of the indigenous inhabitants. In contrast, Alfred Russel Wallace, a well-traveled naturalist and a contemporary of Bates, observed that in Indonesia's Aru archipelago humanity and nature created the conditions for wealth, as exemplified by the region's thriving market in tropical forest products. To Wallace, the diversity of human beings from across the geographical region interacting in carefully scripted exchange relations made for a jubilant display of productivity both of people and of nature.[14] Wallace understood the dynamism between people and nature, but as the biological sciences became increasingly compartmentalized in the twentieth century, such perspectives were seldom seen, and representations of indigenous peoples increasingly cast them as one-dimensional disrupters of ecological equilibrium.[15]

The work of Wallace, Bates, and Humboldt illuminated for their readers fascinating new natural vistas and tropical resources valuable to Europe's growing industrial complex. As Europe colonized Africa and Asia, colonial states sponsored scientific research projects to take inventory of the economic potential of their respective colonies' natural wealth. Germany, for example, began during the 1890s to call upon its scientists to catalog the flora and fauna in its newly acquired East African colony. Prior to these official projects, explorers and colonial enthusiasts had erroneously assumed the riot of vegetation in East Africa's mountain forests, such as those now contained within the Amani Nature Reserve, to be examples of *Urwald*, or a pristine forest primeval.[16] To assess these claims of economic and botanical promise, German East Africa's colonial government asked Adolph Engler, director of the Berlin Botanical Museum, to examine systematically the lush forests covering the Usambara Mountains, located in close proximity to lands designated for settler coffee plantations.

Engler approached his study by disaggregating the mountains' botany into its constituent parts and reassembling it into differentiated ecological communities. Moreover, Engler correctly felt that the complex Usambara forest flora, although seemingly isolated and relatively small in area, could be connected to the continent's natural history. Once he had determined the forests' botanical composition, he built the analysis into a continental-scale history and geography and developed a theory of African plant evolution and adaptation that explained the dynamism of plant evolution. Engler's evolutionary and decidedly ecological view culminated in his massive inventory of African plants, which was first published in 1910.[17]

Engler's 1894 analysis of Usambara's vegetation undermined the previously held idea of a forest *Urwald*.[18] He improved upon the crude ecological triad that appeared in earlier geographical descriptions—settled regions, grazing lands, and forest—by identifying six separate forest communities. Recalling Humboldt's Andean template, he carefully described how each plant community varied with elevation, temperature ranges, and rainfall patterns, as well as the tendency of the forest types to overlap. Engler thus folded Usambara into his intellectual world of systematic botany, which focused on plant form, function, evolution, migration, and adaptation. He also recognized that these forests had seen human occupation despite their pristine appearance. He paid attention to indigenous peoples, but only insofar as they, as foreigners to the plant community, impinged upon the forest's succession.[19] As sophisticated as Engler's work was, it had limitations. The ethnographies, however cursory, of the early natural historians like Humboldt and Wallace did not find their way into Engler's writing nor into the newly developing field of systematic botany. His example shows how late nineteenth-century biological science became increasingly compartmentalized and as a result developed serious blind spots.

The work of these and many other European natural scientists nevertheless wielded a powerful influence on the future of colonial and postcolonial land use. They influenced how governments and people saw the natural world, and they also provided literal descriptions of the diversity of that world. Their descriptions of exotic wilderness fired the imaginations of naturalists in America as well, and shaped the development of the biological sciences and conservation policies in the United States. Through the twentieth century, the landscapes they described became the sites of battles over conservation versus resource extraction, from the Amazon to Indonesia. In the case of East Usambara, the power that lay behind Engler's clear description of the forests' immense biological diversity would lie largely dormant for almost a century until the emergence of conservation biology, when advocates rediscovered these forests' importance as biological hotspots, and historical catalogs of biodiversity became important allies in the fight to conserve the landscapes.[20]

Bounding and Adjudicating Wilderness Landscapes

The places that colonial governments chose for protection reveal much about the European mindset and show surprising parallels with American wilderness history. Conservation fit colonialism's strong tendencies toward paternalism, authoritarianism, and scientific positivism, which compelled colonial governments to designate resources for exploitation or preservation. Whatever its ultimate purpose, natural resource policy created an atmosphere of disenfranchisement and conflict for local peoples. Wilderness preservation projects not only forced removal of those living there, but also the elimination of their history as written on the land.

While nineteenth-century naturalists had identified potential wildernesses, it remained for twentieth-century colonial governments to set aside these regions for preservation. Colonial governments in Africa, for example, divided their territories into legible landscapes—agricultural, forest, pasture, wilderness—and assigned bureaucratic agencies (such as the forest service, game commission, veterinary service, pasture research office, and agricultural service) to manage them. Fears of environmental degradation, and therefore the diminution of the colony's wealth, drove governments to seek ecological balance through the conservation of soils, trees, grasses, and wildlife. At the same time, the administrators responsible for land management tried through authoritarian means to restrict indigenous social and economic life to particular places and eliminate it in others. In many cases, ignorance of ecological, social, economic, and historical conditions reduced colonial conservation for African participants to contentious forced-labor projects.

In eastern and southern Africa, publications documenting the hunting exploits of nineteenth-century white explorers first brought the savanna worlds of Serengeti, Ngorongoro, and the South African Transvaal to the European and American publics. Teddy Roosevelt's subsequent hunting expedition to Tanganyika (1909–1910) cemented the imagery of a Pleistocene remnant of wild savanna teeming with mammalian life. Roosevelt's hunting party participated in what had become a ritual obsession of white elites. The members of his expedition shot thirteen thousand animals, many of which ended up as stuffed specimens in European and American natural history museums. Despite the extraordinary numbers of trophies taken by white hunters, worried game officials deemed indigenous *African* hunting as the major threat to game preservation, a notion that by the late 1920s and early 1930s had spurred organizations like the Society for the Preservation of the Wild Fauna of the Empire to convince Tanganyikan officials to save wild nature in a system of national parks reserved for white tourists.[21]

Ideally, colonial parks like Serengeti and Selous would preserve huge expanses of land as primordial, undisturbed, unchanging, and uninhabited wilderness. Historical research, though, has shown clearly that most of these "wilderness" areas had been inhabited, and the so-called wild landscapes had been shaped by the interactions among herding, farming, and the biophysical environment.[22] Prior to colonial rule, herding societies in nineteenth-century East Africa had undergone a series of devastating crises that included internecine warfare and a rinderpest epidemic that destroyed most of their cattle, discussed in more detail below. Suffice to say here that colonial invasion followed upon this depopulation of a pastoral landscape, and what game officials recognized as a natural landscape teeming with wild animals, herding peoples saw as degraded homelands in need of restoration.[23] According to colonial officials in Tanganyika, however, Africans belonged in such places only if they lived in what European preservationists deemed a "natural state," like the other mammalian museum species in the park.[24]

As colonial nations divided their lands into separate spheres for nature and for people, immigrant settlers began to identify nature with nation. Colonists in New Zealand and South Africa saw in their forests and wild animals, respectively, emblems of nationalist mythology. In 1894, New Zealand founded the

Tongariro National Park, a collection of mountain peaks above the tree line, followed a few years later by a national park in the heavily forested fjord lands of South Island. By 1903, the forests that early colonists had considered a nuisance and an impediment to settlement had become fragmented remnants and, in their growing scarcity, symbols of the landscape's former beauty. The early years of the twentieth century saw the national psyche of New Zealand ascribe to indigenous forests heretofore unknown sanctity as places reserved exclusively for the sustenance of New Zealand's national spirit.[25]

In South Africa, the abundant wildlife of the savanna grasslands, rather than trees, animated the landscapes that the state deemed desirable as wilderness parks. As in East Africa, by the early years of the twentieth century, whites hunted almost exclusively for sport, an endeavor they considered to be morally superior to their African counterparts' subsistence hunting. Kruger National Park, founded in 1926 on two former game reserves, became for Afrikaners, the descendants of Dutch colonists who arrived in South Africa in the seventeenth century, a national symbol that they believed reinforced the primacy of their claims to the land over those of British colonists and the indigenous African ethnic groups.[26] Kruger lay on the grasslands of the Transvaal, a sanctuary in Afrikaner national mythology paid for in pioneer blood and maintained through diligent adherence to Christianity and hard work. For Afrikaners, the park was the landscape unsullied, just as the pioneers found it. The parallels with the nature parks of the U.S. West are inescapable.[27]

The nationalist mythology surrounding these landscapes brims with paradox. What early New Zealand colonialists had seen as nuisances were converted by their descendants into national treasures. The descendants of the English colonists, whom Afrikaner nationalists viewed as oppressors, founded Kruger National Park.[28] Kruger's legacy also points to the convoluted logic necessary to maintain wilderness parks. Upon its founding, Kruger Park authorities evicted three thousand Africans. However, when these same officials almost immediately experienced labor shortages, they reversed their policy, forcing the former African residents to either work as park staff or pay rent.[29] The South African park system's treatment of indigenous residents in many ways mirrored the coercive nature of the apartheid state. In Kruger, park officials prohibited resident African laborers from walking along the park's roads lest the visiting tourists (almost all of them white) got the impression that the park was inhabited by any beings other than animals.

At Selous, a huge park in southeastern Tanzania, officials continually forced African relocations within and around the park's borders in order to accommodate migratory elephant populations, which freely preyed upon African crops without regard for park boundaries. Roderick Neumann, who has conducted extensive research on the Tanganyikan colonial state's park policy, argues that the colonial bureaucracy divided parklands into spaces for conservation and spaces for economic "development." For the Africans who lived in what became Selous National Park, this policy meant being continually forced to relocate away from productive valley farming regions to concentrated settlements in less fertile regions in the interest of an expanding elephant population. Colonial resettlement policy was so successful that elephant numbers exploded to the point where between 1931 and 1950 more than thirty thousand of them had to be shot in order to control overpopulation. Selous, "Africa's last wilderness area," now exists on land that was formerly inhabited by farming communities and was crisscrossed by trade routes.[30]

Postcolonial states and powerful interest groups within them have continued to claim wilderness territories at the expense of resident locals. In the case of Tasmania, wilderness debates demonstrate the power of Australia's environmental movement to assert claims to far-flung lands over the complaints of the area's residents, who know the landscape through long-term residence and use. Tasmania, the island state of Australia, historically evoked among Australians images of a hideous and wild wasteland. However, during the 1960s and 1970s, the growing Australian environmental movement identified Tasmania's remote, mountainous western lands and raging rivers as a valuable wilderness. Tasmania's central plateau emerged in the wilderness discourse as a wild, chaotic, and sacred place, an environmental jewel endangered by human overuse in the form of introduced species, roads, and recreational activities. Despite the fact that Tasmanians of European and Aboriginal descent had used the area for generations, the Tasmanian Green party gained enough political clout in the state legislature to press hard for Eden's restoration on the western plateau. Local users understood wilderness to mean disenfranchisement and cultural imperialism imposed by Australian tourists and hiking clubs.[31]

These examples demonstrate the power of the wilderness concept and the diversity of its application in different regional contexts. In all of these cases, an interest group claims a moral imperative to define wilderness, pushes for its adjudication, and then seeks to determine how local groups fit into the new landscape. The impositions come from outside the local area, and those who know the land through labor and experience lose control of landscapes they had a hand in creating. Wilderness advocates then celebrate park designations uncritically, equating their own aesthetic choices with the appreciation of nature: they champion naturalness in the form of remote areas free from signs of human habitation.[32] The other inescapable insight drawn from these examples is how many resonate with earlier sections of this book and American wilderness history—from the idea of "nature's nation," to Muir's sacred groves, to the Yosemite Indians forced to live in a designated village and work for the national park, to the Western pioneers and the frontier sanctified with sweat and blood, to the backwoods hunters who experienced wilderness parks as dispossession, to the urban nature lovers who support distant parks. It appears that more than just a mode of wilderness preservation has been traveling back and forth across the globe.

Historical Readings of Wilderness Landscapes

In many cases, wilderness designation and management have proceeded under the historical assumption that local residents are destroyers. That humankind has degraded environments across the globe is certainly evident in the archaeological record; however, a growing body of anthropological and historical writing argues for a history of ecological change where human societies have also shaped savanna and forest environments into productive landscapes. Environmental control waxed and waned, as did the sustainability of land-use systems; however, the interrelationships between societies and environments invariably remained dynamic.

Conservation biologist Michael Soulé writes passionately in defense of a nature under siege from a humanity ignorant of modern science (everything from evolution to genetics, and logic as well). Against this background of ignorance, Soulé summarizes for his readers the perspective from biological science of a "living nature," a biological world that exists independent of human manipulation, a concept that opposes directly the constructed nature conceptualized by historian William Cronon. Soulé then outlines what he calls an "ideological war on nature," covertly led in large part by "humanists concerned with the emancipation and empowerment of certain social and ethnic groups." To Soulé, humanists and social scientists cannot speak for nature even though "their views often determine how governments decide to manage wildlands and biodiversity." Finally, Soulé points out the danger of humanistic readings of indigenous history that argue for benign treatments of wildlands, which depend, he says, on their isolation and low population densities. Soulé argues vehemently for a Western-oriented management system based upon the tenets he associates with the biological sciences.[33]

The moralistic tone of many conservation biologists' stance toward preservation and people-free parks is motivated by a concern for biological diversity and the integrity of natural ecosystems. The historical interpretation that buttresses this argument makes the case that wild ecosystems survived only in places where the human inhabitants remained socially unsophisticated and few in number. In contrast, research in environmental history and historical ecology demonstrates the value of understanding humanity's role in the creation, destruction, and renewal of tropical ecosystems. Environmental history can explain the changing ecological diversity of forest and savanna landscapes across a number of geographical and temporal scales. In fact, a multidisciplinary conservation biology that drew on natural history's nineteenth-century tradition of holism and that recognized complex understandings of human history could be exceptionally useful in conserving biodiversity hotspots and the less diverse places that surround them. Recent research on the Amazon basin and the East African savanna demonstrate how history can change the way we understand two of the world's great remaining wilderness areas.

In his summary of the Amazon basin's environmental history, David Cleary points out that few, if any, of the Amazon basin forests escaped use over the last ten thousand years. By the era of nineteenth-century exploration, only one or two centuries had passed since the forest harbored dense populations of farmers.[34] The historical ecologists (the geographers, historians, and anthropologists decried by Soulé) also see a living forest, but one that contained a transformative human society and one capable of creating the conditions for environmental enrichment. As historical landscapes, Amazonian forests have experienced waves of deforestation and regeneration, a constantly shifting mosaic of plants and animals, always in the process of becoming something new.

Laura Rival works among the Huaorani, a foraging group that occupies the heavily forested tributary region of the upper Amazon. In the Huaorani case, Rival's careful analysis of linguistic, botanical, and historical evidence of their material culture reveals a complex ecological understanding of a number of primary and secondary forest types. Some are actively managed as food sources, and others are abandoned yet culturally important botanical and cultural remnants of their ancestors' use.[35] Rival's work follows closely upon the influential studies of William Balée, who has argued similarly that in the Amazonian forests, foraging bands have created useful biotic niches since prehistoric times.[36] The historical research on Latin American forests suggests periods of very dense population and evidence of indigenous management, both intensive and extensive. Resource depletion and enrichment are also evident, as are periods of ecological crisis. In sum, human occupation is part and parcel of the past several millennia of Amazonian forest history, making the "pristine forest" an artifact of historical contingency.[37]

The East African savannas likewise possess a historical narrative that flies in the face of the wilderness lobby's vision of a timeless wonderland under siege from cattle-worshiping pastoralists.[38] The Maasai, in particular, have received a great deal of attention from scholars interested in the environmental history of the East African plains.[39] Historical and archaeological research suggests that there have been two thousand years of pastoralism on East Africa's savannas, stretching back centuries before the relatively recent migration into the area by Maasai pastoralists during the seventeenth and eighteenth centuries.[40] Drawing on already well-established local practices, the Maasai selectively grazed their animals and used fire in order to condition pasture and to delicately tip the balance of disease vectors in favor of cattle and human occupation. The droughts, locust invasions, and disease outbreaks that periodically strike the region required the many ethnic groups to build social relations in order to ease the tensions of these periods of crisis. When cattle population crashes forced Maasai herders to seek refuge with their neighbors, range ecology could suddenly shift to a habitat more suited to wild animals and the insects that carry livestock diseases.

The abundant wildlife described by nineteenth-century explorers and later shot by game enthusiasts appeared on the plains as a result of a series of social, economic, and ecological disasters that struck East Africa during the nineteenth century. The worst of these episodes occurred in the 1890s, when rinderpest, a disease introduced into East Africa by imported cattle, killed the vast majority of cattle across eastern and southern Africa. Predictably, human famine and disease followed the massive livestock deaths. As pastoral societies began a long recovery process, colonial governments seized the opportunity to claim the vast, recently abandoned rangelands for the benefit of settlers and hunters. These ranges would later become the famous wilderness parks of Kenya and Tanzania, like Tsavo, Maasai Mara, Serengeti, and Samburu. Rather than a site of timeless peace and ecological harmony, the long-term history of savanna pastoralism demonstrated continual social, economic, and ecological change, a complex system belying the imagery of the idyllic and balanced savanna so beautifully animated in the opening scenes of Disney's *The Lion King*.[41]

Adding a Cross-Cultural Perspective to Conservation Science

The wilderness concept exhibits an intellectual and cultural history based upon flawed understandings of the dynamic relationships between nature and culture. Even the brief examples presented here demonstrate that the American wilderness model, which values a natural, unsullied, and uninhabited landscape, simply does not hold for many of the world's mountains, forests, and savannas. The incomplete landscape histories that favored wilderness designations by colonial and postcolonial states, moreover, not only disenfranchised local residents, but also denied their role in the production of these valued spaces. However, the research trajectory from the historical sciences that began with eighteenth- and nineteenth-century naturalists and evolved into ecological science has revealed an evolutionary complexity of biologically diverse ecosystems that developed on timescales that far exceed that of human existence. With this in mind, conservation is not only a laudable goal, it is also an ethical responsibility. Unfortunately, the disciplinary gulf between the social sciences and the natural sciences regarding the direction of conservation is reflected all too often on the ground in ongoing conflict. This chapter demonstrates the need for new conceptualizations of conservation that jettison the ahistorical landscape readings so common in the biological sciences. New conservation approaches must also take into account the power of global industrial forces to simplify biologically complex environments through rapid exploitation and demographic change.[42]

Ecology, a young and evolving scientific field, possesses the flexibility to move beyond rigid conservation paradigms. Michael Lewis points out that in India, ecology's evolutionary process includes cross-cultural interaction. In his discussion of the conflicts over the formation of a wilderness park in northern India, Lewis notes the ubiquity of American-trained Indian scientists and the close relationship between the application of American conservation and preservation models and funding dollars from international nongovernmental organizations (NGOs). Despite the financial and institutional power of American science, he also demonstrates that Indian scientists are shaped by their own cultural lives and that they can bring transformative non-Western ideas to the science they practice. As local people practice conservation science, perhaps they will transform conservation by seeing their own human history in the landscape.[43] The application of this emerging hybrid ecology might well enrich conservation models. Unfortunately, at present, the application of conservation science that focuses narrowly on endangered species and habitat protection continues to subordinate local relations among culture, economy, and nature.[44]

Community-based conservation represents another response to the rigidity of the preservationist model. The impetus among NGOs to include local peoples in conservation practice grew out of a drive for social justice, but it also sprang from the realization among conservation advocates that resources immediately outside wilderness parks required sustainable systems in order to protect the people-free core. One estimate claims that 85 percent of the world's protected lands are populated. With such heavy occupation, mass expulsions in the interest of strict preservation are bound to fail, and the ensuing conflict would likely threaten indigenous economic lives, conservation efforts, and the efficacy of unstable states.[45] Community-based conservation is further complicated by the increasingly vague understanding of just who an indigenous person is. In regions of rapid economic and demographic change, or places where long-standing violent conflicts have forced large refugee populations to live in border camps for years, sometimes generations, one finds multiple claims to indigeneity. In this context of instability, community conservation efforts can divide people who vie for political authority and access to the same natural resources.[46]

In such situations, the American wilderness model seems ill suited. Landscapes imagined, occupied, and administered by outsiders do not meet the needs of inhabitants who live in poverty and whose ancestors have experienced the powerful forces of Western exploitation. There is, nonetheless, strong evidence of blossoming environmental movements in the world's tropical regions.[47] On the Indonesian island of Aru, where Alfred Russel Wallace found exotic species and idyllic forests, internationally inspired wilderness projects have spurred local activists to elucidate their own vision of their relationship with nature, one that includes a story of subjection and marginalization by global economic forces.[48] As the moral vision of northern preservationists collides with the calls for justice emerging from the south, wilderness lands will continue to be sites of conflict. While the fight over these spaces continues, the question of what is to happen to nature inside and outside of them remains open.

Notes

1. For a description and history of the Amani Nature Reserve, see http://www.easternarc.org/eucamp/amanigazetted.html.

2. Zephania Ubwani, "Gold Miners Threat to Amani Reserve," *Guardian* (March 31, 2004), available at http://ipp.co.tz/ipp/guardian/2004/03/31/8015.html; Adbud-Aziz, "Gov[ernmen]t to Flush Out Illegal Miners in Tanga," *Guardian* (April 3, 2004), available at http://ipp.co.tz/ipp/guardian/2004/04/03/8190.html; Nike Doggart, Adrian Kahemela, and Phillipo Mbaga, "Gold Mining Threatens the Forests of the Eastern Arc," *Arc Journal* 16 (March 2004): 2.

3. Christopher Conte, *Highland Sanctuary: Environment and History in Tanzania's Usambara Mountains* (Athens: Ohio University Press, 2004), ch. 4.

4. Roderick Nash, *Wilderness and the American Mind*, 4th ed. (New Haven, Conn.: Yale University Press, 2001), 108.

5. Nash, *Wilderness and the American Mind*, 351–58; Stan Stevens, "The Legacy of Yellowstone," in *Conservation through Cultural Survival: Indigenous People and Protected Areas*, ed. Stan Stevens (Washington, D.C.: Island, 1997).

6. Alexander Adams, ed., *First World Conference on National Parks* (Washington, D.C.: U.S. Department of the Interior, 1962); Michael Lewis, *Inventing Global Ecology: Tracking the Biodiversity Ideal in India, 1947–1997* (Athens: Ohio University Press, 2004), 213–14.

7. Gary Paul Nabhan, "Cultural Parallax in Viewing North American Habitats," in *Reinventing Nature? Responses to Postmodern Deconstruction*, ed. Michael Soulé and Gary Lease (Washington, D.C.: Island, 1995), 91. Similarly, on the Columbia River, see Richard White, *The Organic Machine: The Remaking of the Columbia River* (New York: Hill and Wang, 1995), ch. 1.

8. See Gary Nabhan, *The Desert Smells Like Rain: A Naturalist in O'Odham Country* (Tucson: University of Arizona Press, 2002); and *Cultures of Habitat: On Nature, Culture and Story* (Tucson, Ariz.: Treasure Chest, 1998).

9. I include Latin America in this colonial world as indigenous peoples and lands continued to experience conquest and colonization after nations gained independence from Europe.

10. Roderick Neumann, "Africa's 'Last Wilderness': Reordering Space for Political and Economic Control in Colonial Tanzania," *Africa* 71 (2001): 641–43; Paul Greenough and Anna Lowenhaupt Tsing, introduction to their edited collection, *Nature in the Global South: Environmental Projects in South and Southeast Asia* (Durham, N.C.: Duke University Press, 2003), 15; David Cleary, "Towards an Environmental History of the Amazon: From Prehistory to the Nineteenth Century," *Latin American Research Review* 36, no. 2 (2002): 71.

11. For an American example, see William DeBuys, ed., *Seeing Things Whole: The Essential John Wesley Powell* (Washington, D.C.: Island, 2001).

12. For Humboldt's influence upon Muir, see John Muir, *The Cruise of the Corwin* (New York: Houghton Mifflin, 1917).

13. Michael Dettelback, "Humboldtian Science," in *Cultures of Natural History*, ed. N. Jardine, J. A. Secord, and E. C. Spary (Cambridge: Cambridge University Press, 1996), 291.

14. Charles Zerner, "Dividing Lines: Nature, Culture and Commerce in Indonesia's Aru Islands, 1856–1997," in Greenough and Tsing, *Nature in the Global South*, 48–51.

15. For a West African rainforest example, see James Fairhead and Melissa Leach, *Misreading the African Landscape: Society and Ecology in a Forest-Savanna Mosaic* (Cambridge: Cambridge University Press, 1996), ch. 3.

16. For examples, see Oscar Baumann, *Usambara und seine Nachbargebiete: Allgemeine Darstellung des nordöstlichen Deutsch-Ostafrika und seiner Bewohner auf Grund einer im Auftrage der Deutsch-Ostafrikanischen Gesellschaft im Jahre 1890 ausgeführten Reise* (Berlin: Dietrich Reimer, 1891).

17. Adolph Engler, "Die Pflanzenwelt Afrikas insbesondere seiner tropischen Gebiete: Grundzüge der Pflanzenverbreitung in Afrika und die Charakpflanzen Afrikas," in *Die Vegetation der erde*, ed. A. Engler and O. Drude 9, 5 (1:1) (Leipzig, 1925), 226–345.

18. Engler, "Über die Gliederung der Vegetation von Usambara und der angrenzenden Gebiete," in *Abhandlungen der Königlichen Akademie der Wissenschaften zu Berlin*, vol. 1, *Physikalische Abhandlungen* (Berlin: Verlag der Akademie der Wissenschaften, 1894), 43–75.

19. Engler, "Gliederung," 48. Africans employed long-fallow and intercropping techniques using bananas and maize, which led observers to make the mistaken claim of wild banana plants occurring in *Urwald* zones when they were actually seeing the remnants of banana plantings in forest fallows.

20. Other scientists also realized Usambara's biological richness. See Reginald Moreau, especially, "A Synecological Study of Usambara, Tanganyika Territory, with Particular Reference to Birds," *Journal of Ecology* 23 (1935): 1–43.

21. John MacKenzie, *The Empire of Nature: Hunting, Conservation and British Imperialism* (Manchester, England: University of Manchester Press, 1988); Paul Jepson and Robert J. Whittaker, "Histories of Protected Areas: Internationalisation of Conservationist

Values and Their Adoption in the Netherlands Indies (Indonesia)," *Environment and History* 8 (2002): 131.

22. Neumann, "Africa's 'Last Wilderness'," 641–65.

23. Richard Waller, "Emutai: Crisis and Response in Maasailand 1883–1902," in *The Ecology of Survival*, ed. Douglas Johnson and David Anderson (Boulder, Colo.: Westview, 1988), 73–114.

24. Roderick Neumann, "Ways of Seeing Africa: Colonial Recasting of African Society and Landscape in Serengeti National Park," *Ecumene* 2 (1994): 154.

25. Paul Starr, "Native Forest and the Rise of Preservation in New Zealand (1903–1913)," *Environment and History* 8 (2002): 277.

26. Jane Carruthers, "Nationhood and National Parks: Comparative Examples from the Post-Imperial Experience," in *Ecology and Empire: Environmental History of Settler Societies*, ed. Tom Griffiths and Libby Robin (Seattle: University of Washington Press, 1997), 129.

27. For a comparison of South Africa and Alaska, see William Beinart and Peter Coates, *Environment and History: The Taming of Nature in the USA and South Africa* (New York: Routledge, 1995), ch. 5; and Howard Lamar and Leonard Thompson, eds., *The Frontier in History: North America and Southern Africa Compared* (New Haven, Conn.: Yale University Press, 1981).

28. William Adams, "Nature and the Colonial Mind," in *Decolonizing Nature: Strategies for Conservation in a Post-Colonial Era*, ed. William M. Adams and Martin Mulligan (London: Earthscan, 2003), 41.

29. Carruthers, "Nationhood and National Parks," 129.

30. Neumann, "Africa's 'Last Wilderness'," 645–46.

31. Simon Cubit, "Tournaments of Value: Horses, Wilderness, and the Tasmanian Central Plateau," *Environmental History* 6, no. 3 (2001): 397–405.

32. Cubit, "Tournaments of Value," 406.

33. Soulé, "The Social Siege of Nature," in Soulé and Lease, *Reinventing Nature?* 146, 161–62. For a similar view, see John Terborgh, "The Fate of Tropical Forests: A Matter of Stewardship," *Conservation Biology* 14, no. 5 (2000): 1358–61; Jepson and Whittaker, "Histories of Protected Areas," 129–72; Kent Redford and Steven E. Sanderson, "Extracting Humans from Nature," *Conservation Biology* 14, no. 5 (2000): 1362–64.

34. Cleary, "Towards an Environmental History of the Amazon," 69.

35. Laura Rival, "Domestication as a Historical and Symbolic Process: Wild Gardens and Cultivated Forests in the Ecuadorian Amazon," in *Advances in Historical Ecology*, ed. William Balée (New York: Columbia University Press, 1998), 232–50.

36. William Balée, "Indigenous Adaptation to Amazonian Palm Forests," *Principes* 32 (1988): 47–54; "People of the Fallow: A Historical Ecology of Foraging in Lowland South America," in *Conservation of Neotropical Forests: Working from Traditional Resource Use*, ed. K. Redford and C. Padoch (New York: Columbia University Press, 1992), 35–57.

37. Stephan Schwartzman, Daniel Nepstad, and Adriama Moreira, "Arguing Tropical Forest Conservation: People versus Parks," *Conservation Biology* 14, no. 5 (2000): 1371.

38. Dan Brockington, *Fortress Conservation: The Preservation of the Mkomazi Game Reserve* (Oxford: Currey, 2002), 125–26.

39. See works by Richard Waller, including, "Emutai"; "Tsetse Fly in Western Narok, Kenya," *Journal of African History* 31 (1990): 81–101; and "Ecology, Migration, and Expansion in East Africa," *African Affairs* 84 (July 1985): 347–70; and Vigdis Broch-Due, "A Proper Cultivation of Peoples: The Colonial Reconfiguration of Pastoral Tribes and Places in Kenya," in *Producing Nature and Poverty in Africa*, ed. Vigdis Broch-Due and Richard A. Schroeder (Stockholm: Nordic Africa Institute, 2000).

40. John Sutton, *A Thousand Years of East Africa* (Nairobi: British Institute in Eastern Africa, 1990).

41. On this film, see William M. Adams, "Nature and the Colonial Mind," in Adams and Mulligan, *Decolonizing Nature*, 16–17.

42. John R. McNeill, *Something New under the Sun: An Environmental History of the Twentieth Century World* (New York: Norton, 2000).

43. Lewis questions, though, whether urban Indian scientists share more with American peers than with rural Indians. Many Indian ecologists have been at the forefront of governmental attempts to dispossess villagers living in or near national parks. Lewis, *Inventing Global Ecology*.

44. Zerner, "Dividing Lines," 56.

45. Marcus Colchester, "Self-Determination or Environmental Determinism for Indigenous Peoples in Tropical Forest Conservation," *Conservation Biology* 14, no. 5 (2000): 1365.

46. Brockington, *Fortress Conservation*, 115.

47. Greenough and Tsing, introduction to *Nature in the Global South*, 4–5.

48. Zerner, "Dividing Lines," 70. For additional examples of "southern" environmental movements, see Ramachandra Guha, *Environmentalism: A Global History* (New York: Addison, Wesley Longman, 2000), ch. 6.

Fourteen

The Politics of Modern Wilderness

James Morton Turner

"Alaska is our last chance to do it right the first time."[1] Those words launched the most sustained wilderness campaign of the twentieth century. In 1976, the Sierra Club, the Wilderness Society, and other national organizations joined together with Alaskan wilderness groups under the banner of the Alaska Coalition. With the Sierra Club in the lead, the Alaska Coalition took dozens of national advocacy organizations and 1,500 local affiliates and created a well-oiled machine. Its cogs and gears linked the nation's growing environmental constituency with an impressively professional lobbying campaign in Washington, D.C. The Alaska Coalition meshed the old and the new in wilderness advocacy: it joined the citizen activism that had helped to secure the Wilderness Act of 1964 with a new professionalism important to the modern American environmental movement.

As this sophisticated and enthusiastic Alaska campaign gained momentum nationwide, it also galvanized a growing environmental opposition. "Never before have such a small group of well-financed, wild-eyed extremists had the power [of] the Alaska Coalition," complained one opposition group.[2] The state of Alaska and allied industries argued that locking up Alaska's public lands in parks and wilderness areas would limit access to the state's vast petroleum reserves, unlogged forests, and mineral lodes. Another group, calling itself the REAL Alaska Coalition, represented Alaska's sport hunters, off-road vehicle users, and outdoor guides. It argued that parks and wilderness areas would limit opportunities for hunting and motorized recreation. A populist streak coursed through the environmental opposition, aligning citizens, sportsmen, and hunters with the natural resource industries. One opponent scraped "Sierra Club Go to Hell" in the Alaskan tundra with a bulldozer; the earthen banner was so large it could be read from airplanes.[3]

The Alaska lands debate became a bitter political stand-off that culminated in the Alaska National Interest Lands Conservation Act of 1980. Despite major concessions to the opposition, environmentalists succeeded in protecting nearly one-third of Alaska as national parks, preserves, and wildlife refuges. Most important, they succeeded in designating 56 million acres of those lands as federally protected wilderness. In many respects, the Alaska lands act marked a crowning achievement for the American wilderness movement. It doubled the size of the national park and wildlife refuge systems, and it expanded the size of the federal wilderness system fivefold. Roderick Nash summed it up as the "greatest single instance of wilderness preservation in world history."[4] In retrospect, the campaign for Alaska was exceptional, but not for its scope, the grandeur of the landscape, or the natural resources at stake alone. The Alaska campaign was exceptional because it formed a pivotal moment for American environmental politics. No longer could wilderness be described, as the Wilderness Society optimistically suggested in 1979, as American "as motherhood and apple pie."[5]

The central argument of this chapter is that after the Wilderness Act became law in 1964, the idea of wilderness occupied a central, and increasingly contested, place in American environmental politics. How one defined wilderness had real implications for which lands might be included in the national wilderness preservation system. Between 1964 and 2006, wilderness advocates succeeded in expanding the wilderness system from 9.1 million to more than 106 million acres of land. Those successes, however, drew a growing opposition which viewed wilderness, and its proponents, as threatening progress, representative of an inefficient federal government, and unconcerned with individual liberties. Ironically, as wilderness advocates succeeded in expanding the wilderness system, more radical wilderness advocates narrowed their ideas of what wilderness might mean. Drawing on new scientific concerns for protecting biodiversity and consumer-inspired recreation practices, some wilderness enthusiasts idealized wilderness as a truly pristine landscape. These changing political debates over the meaning of wilderness—engaged in by wilderness advocates, scientists, radicals, and their opponents—have combined to make wilderness one of the most divisive concepts in contemporary American environmental thought and politics.

A Pragmatic Wilderness Movement

In 1964, the Wilderness Act launched three wilderness reviews of lands managed by the U.S. Forest Service, the National Park Service, and the U.S. Fish and Wildlife Service. These initial wilderness reviews were not a land grab; they applied only to federal lands already protected by these agencies. Due to a compromise with opponents, however, adding a wilderness area to the wilderness system, then as now, required congressional approval. Whether a 3,000-acre addition or a 6 million–acre addition, every proposal had to follow the same legislative process: field studies, public hearings, agency review, a presidential recommendation, and finally, congressional consideration and legislation (such as the Alaska lands act in 1980). It was a daunting prospect. In 1964, the Sierra Club's executive director, David Brower, warned that implementing the Wilderness Act was "likely to require an enormous amount of constant advocacy in the decade ahead."[6] It was these wilderness reviews that made wilderness a topic of public and, therefore, political debate at the local, regional, and national levels.

Some scholars have suggested, implicitly, that the Wilderness Act settled the value of wilderness for American environmentalism. The law's language and spirit—so carefully crafted by Howard Zahniser—resonated with a broad sweep of American wilderness thought, which included John Muir's temple of nature, Frederick Jackson Turner's nationalistic frontier, Bob Marshall's solitary refuge, and Aldo Leopold's primitive retreat and ecological laboratory. But as the new wilderness reviews began, it became evident that the Wilderness Act raised as many questions about wilderness as it resolved. Translating those broad conceptions of what wilderness means into a system of protected wilderness areas made "wilderness" a popularly contested idea, even among wilderness advocates. Notably, in the 1960s, the Sierra Club and the Wilderness Society did not advance wilderness as a countercultural ideal, envision it as a system of protected ecosystems, nor aim to reform management of the nation's public lands—those ambitions would develop in time. Instead, they carefully formulated a wilderness strategy that avoided extremism, was pragmatic, and was politically popular.

At the heart of the wilderness movement's success was widespread interest in the future of the nation's wildlands. Recreational visits to wilderness jumped from 3 million to 7 million visitor days between 1960 and 1970, and the Sierra Club's membership grew from 15,000 members in 1960 to a political force of more than 100,000 by 1971.[7] Together, the Sierra Club's growing network of chapters and the Washington, D.C.–based Wilderness Society formed an effective institutional framework for mobilizing citizen enthusiasm on behalf of wilderness designations. While most people limited their support to paying annual membership dues, for the thousands who took an active part in wilderness advocacy, these organizations offered an open invitation to participate. They funded local wilderness organizations, encouraged citizen wilderness proposals, organized training programs at the national level, and coordinated citizen lobbying in Congress. The Wilderness Society measured its success not only in the number of wilderness areas protected, but also in the number of citizen leaders trained.[8]

As Harvey pointed out in chapter 11, people found many reasons to appreciate wilderness in the 1960s: some saw it as a recreational escape, others viewed it as a place to hunt, some saw it as an ecological reserve, and some appreciated just knowing that it was protected. How people went about recreating in wilderness began to change in the late 1960s, however. As wilderness areas gained formal protection and gained in popularity, it became evident that they needed to be protected from a growing number of visitors. In peak season, backpackers threatened to overrun the most popular wilderness areas. One forest service employee warned, "The more wilderness is used [for recreation], the less it is a wilderness."[9] In response, backpackers began to focus on minimizing their impact on wilderness. David Brower explained the changing approach to backpacking in 1971: backpacking was about traveling "in harmony with the spirit of wilderness."[10] Another guide suggested the new challenge and reward: it is "quite something . . . to know that you might have harmed a place and that you did not."[11] This approach to outdoor recreation recast the meaning of wilderness for challenge, self-realization, and solitude—ideas important to Bob Marshall and Aldo Leopold—in ways that emphasized minimizing one's impact on wildlands and, by analogy, on the environment as a whole.

New scientific evidence bolstered these concerns for protecting wilderness and the environment as well. Rachel Carson's *Silent Spring* (1962), Paul Ehrlich's *The Population Bomb* (1968), and Barry Commoner's *The Closing Circle* (1970) raised public awareness of the threats posed by a rapidly growing world population and the environmental ills—like polluted rivers, lakes, and air—that faced the modern world.[12] These scientists represented an emerging consensus that described the environment as fragile, interconnected, and fundamentally threatened, basic principles for the emerging environmental movement. In the 1960s, citizens and legislators increasingly saw a new role for the federal government in protecting the nation's environment and its quality of life.

For some wilderness activists, the combined threats of recreational overuse and the larger threats to the environment inspired new ambitions for wilderness protection. Within the Sierra Club, a small group of volunteers began to urge the club to take an uncompromising stand on wilderness in the late 1960s—protecting only the most pristine lands, dedicating wilderness as an ecological preserve, and sharply limiting backpacking. Such strategies were very different from the recreation-oriented wilderness movement of the early 1960s. But, as they saw it, only such a principled stand would truly protect the nation's wilderness and "challenge . . . the basic philosophy of growth and development to which this country adheres."[13] Such sentiments resonated with the radicalism that emerged across America in the late 1960s, as is evident in the student activism of the New Left and the counterculture hippies.

Such ideas would become important a generation later, particularly in the advocacy of Earth First! and the Wildlands Project, but in the late 1960s, the Sierra Club and the Wilderness Society actively distanced themselves from any such radical ideological commitments. In their view, the long-term interests of the wilderness movement depended foremost upon a broad base of popular political support. Challenging the nation's commitment to progress, questioning the recreational value of wilderness, or aligning the wilderness idea with radicalism all threatened, rather than strengthened, their strategy. As a result, the mainstream wilderness movement was careful not to oppose all logging or grazing on the public lands, they made few sweeping claims for protecting wildlife or endangered species, and they avoided challenging the natural resource industries outright. A notable exception to this strategy was David Brower's controversial campaign against dams in the Grand Canyon, which contributed to his ouster from the Sierra Club. As the wilderness advocates knew, protecting wilderness required the active support of legislators in Congress, both Democrats and Republicans. And through the mid-1970s, this pragmatic strategy for protecting the nation's wilderness drew such support.

Three elements were key to this approach to implementing the Wilderness Act. First, wilderness advocates encouraged the recreational use of wilderness areas and argued that the best way to protect the wilderness system from overuse was to protect additional wildlands. "Recreation is one of the named purposes of the Wilderness Act," noted a Sierra Club staffer in 1972. "In fact, it is the first named purpose—out of alphabetical order. I hope this is made plain again and again and again."[14] Second, the wilderness leadership focused on cultivating local political support for wilderness proposals in each state, ensuring that local citizens lobbied the federal agencies and Congress to protect wilderness. As the Wilderness Society's executive director emphasized, success would "hinge on [our] ability to organize and inspire this essential grass roots effort."[15] Finally, the wilderness movement cultivated broad standards for what land could qualify for wilderness designation. Instead of limiting their advocacy to the iconic mountain landscapes of the West, they championed wilderness designations for a broad range of public lands, including small islands, areas close to cities, and restored wildlands in the East and the West.

Despite such careful tactics, the actual expansion of the federal wilderness system was slow. By the end of 1970, after six years of sustained advocacy, Congress had designated only thirty new wilderness areas—places such as the Great Swamp Wilderness in New Jersey, the Pasayten Wilderness in Washington, and the San Rafael Wilderness in California—that added up to just more than a million acres. None of those areas had posed a threat to the natural resource industries or the interests of rural westerners. Indeed, critics described many of the new wilderness areas as "rock and ice," because they protected the most scenic but economically and ecologically least valuable lands at higher elevations. Notwithstanding the slow progress, the *New York Times* noted, "never before has there been such an upsurge of citizen interest and participation in determining the uses to which the American land is devoted."[16] The wilderness movement continued to enjoy mainstream support among the public and bipartisan support in Congress through the 1960s.

New Environmental Strategies for an Old Conservation Issue

The environmental movement captured national attention on April 22, 1970, the inaugural Earth Day. Millions of Americans took part in events, ranging from sit-ins and rallies to stream clean-ups and hikes.[17] In the wake of the turbulent social protests of the 1960s, some observers hoped that the environment would be the issue to unify the nation. As *Time* magazine observed, the environment "attracts the concern of the young and the old, farmers, city dwellers, and suburban housewives, scientists, industrialists, and blue-collar workers."[18] The environment was not the issue of extremists or the counterculture. At least for a moment, one commentator observed, "the environment is . . . as sacred as motherhood."[19] As an oil blowout spewed crude oil along the California coast, a polluted river caught fire in Ohio, and pollution sapped Lake Erie of life, the environmental movement focused national attention on a range of threats to the nation's air, water, wildlife, and human health.

The early 1970s marked the most sustained period of environmental reform in American history. The Nixon administration, reacting to public concern, established the Environmental Protection Agency, and by the mid-1970s a proactive Congress passed or amended a wide range of environmental legislation, including the National Environmental Policy Act, the Clean Air Act, and the Endangered Species Act.[20] In these years, a new generation of environmental organizations, such as the Natural Resources Defense Council and the Environmental Defense Fund, organized to address the new environmental issues. Amid a growing concern with environmental quality and pollution, some activists and historians asked if the wilderness movement would be left behind. Michael McCloskey, the Sierra Club's new executive director, warned that in the "context of the new environmental movement, wilderness preservation appears to many as parochial and old-fashioned."[21]

The wilderness movement, instead of being left behind by the emerging environmental movement in the 1970s, repositioned itself to draw on the movement's new tools—legislative, legal, and scientific—to advance the campaign for wilderness. Most important to this strategy was the National Environmental Policy Act of 1969 (NEPA), which required the federal government to consider the environmental effects of government-sponsored activities in "environmental impact statements." Using this law, the wilderness movement made three strategic advances in the early 1970s that dramatically expanded the potential for wilderness protection nationwide. First, the Wilderness Society intervened in legislation to distribute lands to Alaska's natives and plans to build a trans-Alaska pipeline to carry oil from northern Alaska to the southern port of Valdez. Second, the Sierra Club filed suit against the forest service over its plan to open up roadless areas in the national forests for logging. Third, the Sierra Club and the Wilderness Society cooperated in advancing new legislation to guide the management of the public lands controlled by the Bureau of Land Management (BLM).

At the heart of each of these campaigns was the rationale, legislated by NEPA, that the government must consider the environmental consequences of and potential alternatives to its actions. The wilderness advocates used that principle to achieve what the Wilderness Act had not: a wilderness review of almost all of the nation's remaining federal wildlands. Where the Wilderness Act initially promised to protect 50 million acres of land, by the mid-1970s wilderness advocates hoped that they might one day protect 100 million acres of land or more. Indeed, the federal government's new wilderness reviews included more than 250 million acres of land. These new reviews, however, directly challenged the natural resource industries' and rural westerners' access to the public lands. Industry executives began complaining of the "pressures of the extreme environmentalists" and argued that such public involvement in resource management led to "emotional rather than objective" decisions.[22] The wilderness movement had succeeded in hitching its future to the larger environmental movement. That strategy, however, set the stage for a backlash against the wilderness movement.

The Polarization of Wilderness Politics

The bipartisan and popular consensus behind environmental reform began to fall apart in the mid-1970s. In part, this can be blamed on concerns about a weakening economy and rising energy prices, as well as the surprising successes of the environmental movement. The political backlash against the environmental movement, however, was not a straightforward reaction to either. Rather, the environmental opposition and the environmental movement itself reacted to broader changes in environmentalism and the role of the federal government in environmental protection in the late 1970s and early 1980s. At times, they reacted to these changes in surprisingly similar ways. A well-organized and aggressive environmental opposition, known as the Sagebrush Rebellion, and a more idealistic and radical current of environmental activism—best represented by Earth First!—emerged in parallel in these years. These dual critiques of environmental reform established new boundaries for American environmental thought and politics that persist to the present day.

The election of environmentally minded President Jimmy Carter seemed to promise a rosy future for the wilderness movement. With the new wilderness reviews under way for Alaska, the national forest roadless areas, and, soon, the Bureau of Land Management lands, the cards seemed stacked in the wilderness movement's favor: wilderness was a national priority. For many Americans, the campaign for Alaska appeared to be the focal point of the wilderness movement's activities in the late 1970s. But across the lower forty-eight, the wilderness reviews of the national forest roadless areas formed an equally important arena of debate. Yet, unlike Alaska, which was a highly political contest from the start, the forest service pitched its wilderness review as a rational, comprehensive, and apolitical assessment of the nation's remaining 60 million acres of wild national forest land.[23]

The forest service's new wilderness review was representative of the bureaucratic approach to environmental decision making that had become increasingly common since NEPA became law. After an earlier national forest review failed in 1973, on grounds that it did not satisfy the requirements of NEPA, the forest service launched the second Roadless Area Review and Evaluation (RARE II) in 1977. The agency's goal was to assess the scientific, economic, and public interest in more than 3,000 roadless areas in the national forests in less than two years. RARE II was an extraordinary undertaking: it involved hundreds of local meetings nationwide, a quarter of a million public comments, and detailed resource assessments and economic analyses of the national forests. Where earlier wilderness reviews and proposals had proceeded at a small scale, considering a few potential wilderness areas at a time, RARE II examined all of the remaining roadless areas at once. That strategy gave new emphasis to an abstract language of natural resource analysis rooted in economics and environmental analysis.

In 1979, the Carter administration announced the forest service's final recommendations for the 62 million acres of national forest roadless areas: 36 million acres should be released as nonwilderness, 10.6 million acres should be left in further planning, and 15.4 million acres should be designated as wilderness.[24] If the forest service had achieved its goal with RARE II, the final recommendations would have provided a clear outline for congressional action and resolved debate over the national forest roadless areas. But RARE II fell short of the agency's, the wilderness advocates', and industries' expectations. The wilderness movement viewed it as insufficient; they had hoped the agency would recommend wilderness protection for 36 million acres of roadless areas.[25] The timber industry, in private, considered the review a success, but worried that the final recommendations would be challenged in court. Ultimately, the review collapsed into a highly politicized legal and legislative stalemate that would not break until 1984 and that remains a point of contention to the present day.

RARE II challenged the wilderness movement's earlier approach to wilderness advocacy. The highly technical review process diminished the influence of citizen volunteers. While such citizen involvement was encouraged by the agency and wilderness groups, monitoring RARE II required careful cooperation and, most important, consistency nationwide: efforts to advance any one wilderness proposal had complex effects on other wilderness proposals at the state and national levels. This centralization of RARE II helped to consolidate an increasingly influential and tight-knit group of professional wilderness lobbyists in Washington, D.C., from the Wilderness Society and the Sierra Club. To keep pace with such environmental reviews, these organizations hired scientists, economists, and other professionals, and they embraced a more professional organizational structure.[26] The Wilderness Society, in particular, underwent a dramatic organizational restructuring starting in 1978, shifting its priorities away from citizen organizing and toward professional advocacy. As its new executive director explained, his strategy was "more responsibility and less stridency; more professionalism and less emotionalism; more dialogue and less diatribe."[27]

That prescription, however, was as much a reaction to changes already under way in environmental politics as it was a forward-looking vision for the future. As the new wilderness reviews got under way, the *Conservative Digest* summed up a growing opinion of the wilderness reviews: "The Big Federal Land Grab."[28] In the late 1970s, industries became increasingly aggressive and savvy in their opposition to the wilderness movement. They lobbied Congress and the Carter administration, and they adopted the political strategies that had worked well for the wilderness movement, such as organizing local citizens. During RARE II, protesters picketed forest service offices, displaying signs reading: "[We] need more timber sales—not wilderness"; "Stop the Sierra Club"; "We can't make a living by hiking."[29] RARE II drew a growing opposition, similar to that in Alaska, which was led by the natural resource industries and supported by an alliance of rural employees, hunters, ranchers, and off-road vehicle enthusiasts. Rooted in the rural West, angry over a slowing economy, and frustrated with environmental regulations, these opponents formed but a part of a broader, long-building conservative outcry that swept Ronald Reagan into the White House in 1980. Gathering under the banner of the Sagebrush Rebellion, these groups invoked a powerful language of states' rights to challenge an expanding federal government.[30]

For the Wilderness Society's and the Sierra Club's leadership, the challenge was not just countering the forest service's limited wilderness recommendations or the growing political opposition of the Sagebrush Rebellion. It was also addressing an emerging current of environmental radicalism that threatened to fragment the wilderness movement's unity. The sharpest challenge to the mainstream organizations came from Earth First! Founded by five wilderness activists, two of whom were former Wilderness Society organizers, Earth First! positioned itself as the antithesis of the newly restructured and professional Wilderness Society. As Dave Foreman, an Earth First! cofounder and its most visible leader, later explained: "It was time for a new joker in the deck: a militant, uncompromising group unafraid to say what needed to be said or to back it up with stronger actions than the established organizations were willing to take."[31] Earth First! gained notoriety for encouraging covert "monkey-wrenching" activities that sabotaged development activities, such as logging and road building, and for more overt activities, such as picketing government offices and blockading logging roads—all in the name of fighting for wilderness. To Earth First! fighting for wilderness was not just a political process, but "a battle for life itself, for the continued flow of evolution."[32]

These powerful currents of protest reshaped the landscape of environmental politics and reflected broader shifts in national politics in the early 1980s. The Reagan administration catered to the Sagebrush Rebellion and launched an antienvironmental campaign, led most prominently by Secretary of the Interior James Watt, which aimed to undermine the wilderness system and roll back the nation's environmental regulations. This new streak of Republican antienvironmentalism forced compromise on the Alaska legislation, stalled progress on wilderness designations in the national forests, and threatened to stop the reviews for the BLM lands, which were just beginning in the early 1980s. This broad attack on the public lands energized the environmental community: membership in environmental groups swelled, a petition against James Watt drew a million signatures, and wilderness advocates continued to advance a campaign to address the shortcomings of RARE II. Those latter efforts culminated in 1984, when the Wilderness Society and the Sierra Club successfully piloted eighteen wilderness bills protecting 8 million acres of the national forest roadless areas through Congress despite strong opposition. Earth First! however, dismissed those bills as political compromises, dubbing them the "Wilderness Destruction Acts of 1984," since they released millions of acres of potential roadless areas for logging and development.[33]

Radical sentiments for and against wilderness emerged in parallel in the late 1970s and early 1980s. While the political aims of the Sagebrush Rebellion and Earth First! were diametrically opposed, their antifederalist inspiration and rhetoric were rooted in a shared set of frustrations with the institutionalization of wilderness politics and environmentalism in the 1970s. The Sagebrush rebels embraced a grassroots rebellion against "overgovernment," "Big Brother," and "extreme environmentalists," all based in Washington, D.C. In their view, the wilderness system, rather than representing the nation's patriotic heritage, represented the excesses of the federal government. Earth First! envisioned the wilderness system as an ecological resource that could not be compromised in the political arena and sought to liberate the wilderness movement from the bureaucracy of Washington, D.C. The radicalism of both Earth First! and the Sagebrush Rebellion established new endpoints on the ideological spectrum of American environmental and political thought. In the 1980s, the rhetoric of environmental politics increasingly tended toward those extremes, undermining the pragmatic approach to wilderness that had been at the center of the wilderness movement's legislative successes in the 1970s.

New Visions for the American Wilderness

Concerns that the wilderness movement had lost focus swept through its ranks in the mid-1980s: had it become too professional? too focused on Washington, D.C.? ill prepared to cultivate the citizen support that formed the backbone of its early successes? Ernie Dickerman, a long-time wilderness advocate, offered his biting assessment of the Wilderness Society in 1987: "The Wilderness Society has become one more cozy, private, Washington bureaucracy, happily spinning its wheels, free from the dirt and sweat of citizen contact, self-satisfied with its frequent, erudite publications."[34] To some observers, it seemed that the national wilderness organizations were bewildered by a growing opposition and unable to cultivate national, bipartisan support for wilderness protection. In the late 1980s, the mainstream wilderness movement challenged the economic underpinnings of federal public lands policy in the American West, and more radical wilderness advocates began to embrace idealistic visions for wilderness protection, best represented by institutions such as the Wildlands Project and Leave No Trace.

Many western states watched traditional extractive industries, such as grazing, logging, and mining, sputter in the face of overproduction, international competition, and domestic environmental regulations in the 1980s. As the economic future of the rural West became more uncertain, wilderness politics hardened into a bitter stalemate pitting jobs against wilderness, rural communities against urban environmentalists, and Republicans against Democrats—a pattern that reflected the tenor of environmental politics nationwide. These challenges were evident in the BLM wilderness review, which began in the early 1980s. Efforts to protect the BLM lands drew sharp opposition, first from the Sagebrush Rebellion and then from the Wise Use movement in the late 1980s. Wise Use, which emerged out of the Sagebrush Rebellion, claimed to speak for workers, families, and communities in the rural West that were threatened by environmental laws that weakened the region's economy and encroached on private property rights.

In fact, groups like the Wilderness Society and the Sierra Club *had* begun to challenge the traditional economy of the American West. In particular, the Wilderness Society drew on its professional resources—the economists, ecologists, and natural resource specialists it employed—to challenge the basic assumptions that underlay federal land management. Starting in the mid-1980s, the Wilderness Society published a series of technical reports which questioned the economics of resource development on the public lands. They argued that the federal government was giving away the nation's public natural resources to private interests, such as miners, loggers, and ranchers, at rates that were far below market value. Mining companies gained title to public lands under antiquated mining laws for $5 per acre. Ranchers paid approximately one-fifth of market value to graze cattle. And the forest service sold public timber at a net loss nationwide. For instance, a 1984 Wilderness Society study of the 155 national forests revealed that 75 national forests failed to recover even half of the expenses incurred in running the timber programs.[35] In 1986, the *Washington Post* editorialized, "In a year when so many other forms of federal support are in jeopardy"—in reference to the Reagan administration's efforts to cut social welfare and other government programs—"there is no excuse for exempting these."[36]

The Wilderness Society worked in concert with other environmental groups to raise a broad set of questions about the economic basis and ecological implications of public lands management. Where in the 1970s, the wilderness movement had been careful to avoid challenging the natural resource industries' access to the public lands outright, starting in the mid-1980s, the wilderness movement questioned the most basic economic assumptions used to justify development of the public lands. Drawing on the scientific principles of conservation biology, the Wilderness Society argued that, if such policies did not make economic sense, it was clear that they did not make ecological sense either. Indeed, overgrazing on the public lands, clear-cutting the national forests, and the consequences of mining all threatened the nation's wildlands and the biodiversity they supported. These arguments became a focal point of debate in the Pacific Northwest in the late 1980s. There, loggers and environmentalists divided over the future of the region's remaining old-growth forests, the unique ecosystems they represented, and the fate of a threatened species, the spotted owl.

While the mainstream wilderness movement focused on broader questions of public lands management, the most dynamic arena of wilderness advocacy emerged among more radical wilderness activists. A new group—the Wildlands Project—linked the self-described "new conservation movement" with the emerging science of conservation biology. Founded in 1991, the Wildlands Project was a descendant of Earth First! (Dave Foreman helped to found both). But where Earth First! had championed extralegal activities in the defense of wilderness, the Wildlands Project aimed to reshape the nation's vision for wilderness protection. In the 1990s, the Wildlands Project emerged as a hub of activity for a growing network of grassroots wilderness advocates who had grown frustrated with the mainstream wilderness organizations. The Alliance for the Wild Rockies and Preserve Appalachian Wilderness were but two of the dozens of organizations that had begun to promote expansive new wilderness proposals based on the scientific principles of conservation biology and a philosophical commitment to a biocentric world view. In 1993, the Wildlands Project gathered these proposals together into an overarching vision for the future of wilderness in the United States. It proposed protecting one-half of the nation's land as "core reserves" and "inner corridor zones" in the name of protecting biodiversity. *Science* magazine described it as "the most ambitious proposal for land management since the Louisiana Purchase of 1803."[37]

The Wildlands Project's vision of wilderness protection was a powerful alternative to the modest goals for wilderness protection that dominated national wilderness debates in the 1980s. Even as national groups questioned the economics of public lands policy, they avoided large-scale wilderness proposals, for fear of provoking further political backlash. In contrast, the new conservation movement advanced proposals such as the Northern Rockies Ecosystem Protection Act (NREPA). Where the mainstream groups proposed protecting 1 or 2 million acres of national forests in Montana and Idaho near Yellowstone National Park, NREPA proposed protecting upward of 15 million acres.[38] According to the Alliance for the Wild Rockies, only such a grand proposal, establishing a network of interconnected parks and wilderness areas, could support grizzly bears, wolves, and the rest of the region's biodiversity in the long term. Such an ambitious proposal may have been politically unworkable, but as one Wilderness Society staffer warned, if the organization did not embrace the new way of thinking, "we'll deal ourselves out of the discussion."[39] By the mid-1990s, the mainstream wilderness movement began to adopt the language, if not the ambitions, of the new conservation movement. In observation of the thirtieth anniversary of the Wilderness Act, the Sierra Club announced its new "large-scale ecosystem protection program." While the previous generation of wilderness advocates secured the first 100 million acres of wilderness, they explained, "a new generation of wildlands advocates, as passionate as the old-timers, but much more scientifically savvy, has begun going after the next eighty million acres."[40]

As some wilderness advocates gave new emphasis to wilderness as a scientific ideal, others gave wilderness new value as a consumer ideal. Since the 1960s, recreational use of wilderness had continued to grow: recreational visits to the most popular wilderness areas increased, and companies catering to backpackers, such as REI, Patagonia, and Timberland, expanded into multimillion-dollar corporations. This growing commercial interest in wilderness recreation and the work of nonprofit wilderness advocacy began to align in the mid-1970s. "I manage a 22 million dollar outdoor equipment co-op with 524,000 members," explained REI's general manager, James Whittaker, to Congress. "We need wilderness and natural areas for our business."[41] By the early 1990s, REI had provided nearly $1 million to support wilderness advocacy. Timberland Corporation sponsored the Wilderness Society's twenty-fifth anniversary celebration of the Wilderness Act in 1989. And Patagonia provided seed funds to some of the groups that comprised the new conservation movement starting in the 1990s. This growing alliance between wilderness advocates and consumer retailers marked a distinct shift away from the origins of the modern wilderness ideal, which Paul Sutter discussed in chapter 10. Where the Wilderness Society's founders valued wilderness, in part, as a social critique of consumerism in the 1930s, wilderness advocates began to promote wilderness as a consumer ideal in the 1980s.

This changing understanding of wilderness and outdoor recreation was best represented by the formation of Leave No Trace in the early 1990s.[42] This program gathered together the minimal-impact camping techniques pioneered in the 1970s and organized them into a well-advertised set of guidelines for wilderness recreation. With the support of the federal government and wilderness advocacy groups and considerable financial support from the outdoor recreation industry, Leave No Trace became the standard for environmentally friendly outdoor recreation. The most devoted backpackers fluffed the grass on which they had slept, gave up toilet paper rather than burying it, and preferred drinking their own dishwater to pouring it on the ground. No measure seemed too extreme in their efforts to "leave no trace." Despite the success of Leave No Trace in protecting wilderness from recreational overuse, this recreation ethic institutionalized a very specific—and at times expensive—way of appreciating and valuing the nation's wilderness. Unlike earlier wilderness enthusiasts, who prized wilderness as a place to live off the land and experience a primitive hunt—those ideas had been particularly important to Aldo Leopold—Leave No Trace left little room for hunters and rural westerners, who had their own ways of appreciating wilderness.

The Wildlands Project and Leave No Trace represented powerful reformulations of the long-standing scientific and recreational interests in wilderness. In important ways, these approaches to wilderness were complementary. Both institutions cultivated wilderness as an idealized landscape that was meant to be pristine, unpeopled, and ahistoric. Taken together, this idealized recreational wilderness and this ambitious biocentric wilderness both evoked a high set of standards (perhaps impossibly high) for what wilderness might mean. While such sentiments had always formed an important current of wilderness thought, a more pragmatic approach to wilderness had oriented the politics of the mainstream wilderness movement since the 1960s. Starting in the late 1980s, however, this growing current of wilderness fundamentalism began to reorient the mainstream wilderness movement's goals. The Wilderness Society and the Sierra Club adopted strategies that reflected this uncompromising approach to wilderness: putting forward proposals to protect large-scale ecosystems, to ban logging in all national forest roadless areas, and to curtail grazing in wilderness areas and on the public lands. Such proposals gained new political possibility in the early 1990s.

The 1992 election of Bill Clinton raised environmentalists' hopes for a sea tide of change in Washington, D.C. The *Washington Post* reported that the incoming Clinton administration "has set toes tapping in anticipation at groups such as The Wilderness Society and the Sierra Club, and worry beads jiggling at groups that represent miners, cattlemen, irrigators and the wood products industry."[43] Indeed, wilderness advocates expected quick action on a number of wilderness issues, including the stalled BLM wilderness reviews, the national forests in the northern Rockies, and the Arctic National Wildlife Refuge's coastal plain. Despite such high hopes, however, few wilderness bills became law in the 1990s. Most notable was the California Desert Protection Act of 1994—and it was nearly blocked by the Wise Use movement, which protested what it described as the Clinton administration's "war on the West."[44] When the Republicans gained control of the House of Representatives in 1994, enacting wilderness legislation proved to be a near-impossible hurdle. Faced with sharp Republican opposition, wilderness activists shifted their strategies away from generating popular support in Congress and instead focused on pressuring the Clinton administration to take executive action. Those efforts culminated in the protection of 58.5 million acres of national forest roadless areas and 6 million acres of national monument designations in the closing days of the Clinton administration.

The fact that those successes came by way of executive action (which is open to later reversal by future presidents or Congresses—and, indeed, President George W. Bush overturned the roadless designation in 2005) reflected the degree to which the opportunities for legislative compromise in Congress were blocked during the 1990s. Instead of drawing opponents in toward a moderate center, this political landscape pushed advocates out toward the radical margins of environmental politics. During the 1990s, the most radical allies of the Wise Use movement spun off into violence: burning environmentalists in effigy in eastern Oregon; issuing death threats to environmentalists in New Mexico, California, and Washington; threatening federal employees on the job; and even firing shots at hikers in the Utah backcountry.[45] Campaigns of monkey wrenching and other forms of civil disobedience, such as tree sitting and blockades, continued to mark more extreme efforts to protect wilderness. And, of course, the Wildlands Project and its allied new conservation movement charted ever more ambitious visions for the wilderness movement. In this divisive political context, the mainstream wilderness organizations struggled to position wilderness as an issue of national interest. Instead, as the wilderness movement emphasized the value of biodiversity and the importance of backpacking, and as it questioned the economics of the public lands, the wilderness movement increasingly appeared to be the partisans of a special interest.

Conclusion

When the Wilderness Act of 1964 became law, it was rightly regarded as a national affirmation of the value of wildlands for present and future generations of Americans. The *New York Times* described it as a "landmark." Amid the political controversy that entangled wilderness in the 1990s, the value of wilderness to the nation, even to the environmental movement, was no longer so clear. Of course, wilderness advocates had weathered the critiques of the Sagebrush Rebellion and the Wise Use movement since the 1970s, but those could be dismissed as political arguments advanced by those most vested in opposing wilderness. Starting in the late 1980s, however, the scholarly community began to challenge the American wilderness ideal. The most public of these critiques came from the environmental historian William Cronon. In his 1995 essay, "The Trouble with Wilderness," discussed in this volume's introduction, he argued that wilderness had become the "unexamined foundation on which so many of the quasi-religious values of modern environmentalism rest."[46]

In the 1990s, this "great new wilderness debate" suggested that America's peculiar fascination with protecting wilderness had blinded us to broader social and environmental concerns both at home and abroad. Scholars argued that in idealizing wilderness as an unpeopled landscape, we often overlooked the socioeconomic implications of such an ideal and the interests of the peoples who relied on those lands for their welfare, including Native Americans, rural Americans, and indigenous peoples abroad. Scholars warned that to the extent we elevated wilderness as nature—the "true" nature most worth saving—we risked overlooking the environmental consequences of our own activities nearer to home. And other scholars have worried that wilderness issues—such as the fate of the Arctic National Wildlife Refuge—distract us from more pressing environmental issues, such as climate change. Many scholars have suggested that environmentalism needs to focus on approaches to environmental protection that promote a working relationship between society and nature, such as sustainable development and environmental justice, rather than separating people from nature.[47]

The great new wilderness debate was framed as a broad critique of the American wilderness ideal. Yet, for all of the debate's merits in raising questions about the assumptions that underlie the modern wilderness ideal, many scholars overlooked the historical contingency of wilderness in American environmental thought. They traced wilderness as if it were an unchanging idea from John Muir's advocacy in the nineteenth century through the Wildlands Project's ambitious wilderness advocacy at the end of the twentieth century. In many regards, however, the great new wilderness debate marked a specific reaction to the new wilderness fundamentalism of the late 1980s, not a general reaction to the more pragmatic concerns that guided wilderness advocacy during the rest of the twentieth century. Returning wilderness to the contentious stage of environmental history suggests that while wilderness has been informed by overarching ideals, it has also been profoundly shaped by the shifting imperatives of science, recreation, and politics. The wilderness politics of the last forty years of the twentieth century, like those in earlier periods of American history, emerged out of unique circumstances that reflected the changing place of wilderness in American environmental thought and politics. Wilderness has never been an idle concept.

Notes

Several manuscript collections are referenced. The Sierra Club Member Papers, MSS 71/103c, are in the Bancroft Library, University of California, Berkeley (hereafter SCMP). The Sierra Club Pacific Northwest Papers are in the Allen Library, University of Washington, Seattle (SCPNWP). The remaining collections are all in the Conservation Collection, Denver Public Library, Denver, Colorado: Alaska Coalition Papers, CONS89 (ACP); Colorado Environmental Coalition Papers, CONS137 (CECP); Harry B. Crandell Papers, CONS86 (HCP); Wilderness Society Papers, CONS241 (TWSP). Please note that the Wilderness Society Papers were reorganized by the Denver Public Library in 2005, and some reference numbers may have changed.

1. Alaska Coalition, "Alaska Slide Show Script," n.d., ACP, Box 4, folder "Grassroots—slide shows scripts."

2. Citizens for Management of Alaska Lands, "Letter to Dale Philman, Alaska Wilderness Expeditions," August 31, 1978, TWSP, Box 12:300, folder "ANILCA: Correspondence—General Aug–Oct 1978."

3. Alaska Chapter, Sierra Club (hereafter SC), "Alaska Newsletter," October 10, 1978, TWSP, Box 12:300, folder "ANILCA: Correspondence: General Aug–Oct 1978."

4. Roderick Nash, *Wilderness and the American Mind*, 3d ed. (New Haven, Conn.: Yale University Press, 1982), x.

5. William A. Turnage, Wilderness Society (hereafter TWS), "Confidential Special Report on Alaska Lands Legislation," February 27, 1979, Box 3, folder "Conservationists Involvement in Alaska," TWSP.

6. David Brower, "Wilderness and the Constant Advocate," *Sierra Club Bulletin* (September 1964), 3.

7. Nash, *Wilderness and the American Mind*, ix.

8. Stewart M. Brandborg, TWS, "Letter to Members," May 9, 1966, TWSP, Box 7:173, folder "Tennessee: GWMNP, May 1966."

9. William A. Worf, U.S. Forest Service (hereafter USFS), "The Commercial Outfitter and Wilderness," January 3, 1970, Arthur Carhart Wilderness Training Center Archives (Missoula, Mont.), document no. 161.

10. David Brower, ed., *The Sierra Club Wilderness Handbook*, 2d ed. (New York: Sierra Club/Ballantine, 1971), 55.

11. John Hart, *Walking Softly in the Wilderness* (San Francisco, Calif.: Sierra Club Books, 1977), 230.

12. Rachel Carson, *Silent Spring* (Boston: Houghton Mifflin, 1962); Paul Ehrlich, *The Population Bomb* (New York: Ballantine, 1968); Barry Commoner, *The Closing Circle: Nature, Man, and Technology* (New York: Knopf, 1971).

13. Walcott, "Wilderness, Why, What and How?" 1970–, SCMP, Box 222:29, folder "Wilderness Classification Study Committee, n.d."

14. Brock Evans, SC, "Letter to Harry Crandell, TWS," December 15, 1972, HCP, Box 2:5, folder "Wilderness management—guidelines/jurisdiction/classification."

15. Brandborg to "Members," May 9, 1966.

16. "Editorial: The Wilderness Hearings," *New York Times* (January 7, 1967), 24.

17. See Robert Gottlieb, *Forcing the Spring: The Transformation of the American Environmental Movement* (Washington, D.C.: Island, 1993); Hal K. Rothman, *The Greening of a Nation? Environmentalism in the United States since 1945* (New York: Harcourt Brace, 1998); Kirkpatrick Sale, *The Green Revolution: The Environmental Movement, 1962–1992* (New York: HarperCollins, 1993).

18. "Fighting to Save the Earth from Man," *Time* (February 2, 1970), 56–57.

19. Paul O' Neil, "Walter Hickel Is an Endangered Species," *Life* (August 28, 1970), 48A+.

20. J. Brooks Flippen, *Nixon and the Environment* (Albuquerque: University of New Mexico Press, 2000).

21. Michael McCloskey, "Wilderness Movement at the Crossroads," *Pacific Historical Review* 41 (1972): 346–52.

22. Robert O. Anderson, Chairman of the Board, Atlantic Richfield Company, "Letter to President Nixon from Anderson," August 27, 1970, Box 26, John Whitaker Papers, National Archives, College Park, Md.; Wendell B. Branches, Western Wood Products Association, "Letter to Neal Rahm, USFS," July 22, 1970, Box 192b, National Forest Products Association Papers, Forest History Society, Durham, N.C.

23. See Dennis M. Roth, *The Wilderness Movement and the National Forests: 1964–1980* (Washington, D.C.: GPO, 1984).

24. U.S. Forest Service, *Final Environmental Impact Statement: Roadless Area Review and Evaluation* (Washington, D.C.: GPO, 1979).

25. Citizens for America's Endangered Wilderness, "RARE II: The Results," January 1979, CECP, Box 4:30A, folder "Issues: Memos—Wilderness Areas, 1979–1984."

26. See Gottlieb, *Forcing the Spring,* ch. 4.

27. William A. Turnage, TWS, "Letter to Richard King Mellon Foundation," August 8, 1979, TWSP, Box 6, folder "Mellon Foundation Proposal."

28. "The Big Federal Land Grab," *Conservative Digest* (December 1978).

29. "Wilderness Proposals Draw Protest in Macon County," *Asheville Times* (July 12, 1977), 1.

30. R. McGreggor Cawley, *Federal Land, Western Anger: The Sagebrush Rebellion and Environmental Politics* (Lawrence: University Press of Kansas, 1993).

31. Dave Foreman, *Confessions of an Eco-Warrior* (New York: Harmony, 1991), 17.

32. "The Problem," *Earth First! Journal* (December 1981).

33. Bill Devall, "Editorial," *Earth First! Journal* (December 1, 1984).

34. Ernest P. Dickerman, "Letter to Harry Crandell," April 27, 1987, HCP, Box 2:4, folder "Correspondence from the Wilderness Society, 1975–1988."

35. Alaric Sample, Jr., "Below-Cost Timber Sales on the National Forests," (Washington, D.C.: Wilderness Society, 1984), 10.

36. "Editorial: User Fees," *Washington Post* (February 3, 1986), 14.

37. Charles C. Mann and Mark L. Plummer, "The High Cost of Biodiversity," *Science* 260, no. 5116 (June 25, 1993), 1868–71.

38. See "The Last Best Chance," *Sierra* (March 1994), 40+.

39. Craig Gehrke, TWS, "Memo Re: Monthly Report," December 20, 1992, TWSP, Box 19:1, folder "Field Programs, Monthly Reports."

40. Bruce Hamilton, "An Enduring Wilderness?" *Sierra* 79, no. 5 (September 1994): 46–49.

41. James W. Whittaker, REI, "Letter to James Haley, Chairman, House Interior Committee," January 23, 1976, SCPNWP, Box 2678–2, 3, folder "Alpine Lakes, General Correspondence, Ja[nuary] 1976."

42. Jeffrey L. Marion and Scott E. Reid, *Development of the U.S. Leave No Trace Program: An Historical Perspective* (Boulder, Colo.: Leave No Trace, 2001).

43. Tom Kenworthy, "Pragmatic Critic Is Set to Be Interior's Next Landlord," *Washington Post* (January 19, 1993), 9.

44. Tom Kenworthy, "House Passes Desert Protection Measure," *Washington Post* (July 28, 1994), 5.

45. Rothman, *Greening of a Nation?* ch. 8; Jacqueline Switzer, *Green Backlash: The History and Politics of Environmental Opposition in the United States* (Boulder, Colo.: Lynne Rienner, 1997), ch. 8.

46. William Cronon, "The Trouble with Wilderness; or, Getting Back to the Wrong Nature," in *Uncommon Ground: Rethinking the Human Place in Nature,* ed. William Cronon (New York: Norton, 1995), 80.

47. J. Baird Callicott and Michael P. Nelson, eds., *The Great New Wilderness Debate* (Athens: University of Georgia Press, 1998).

Epilogue

Nature, Liberty, and Equality

Donald Worster

The struggle to protect nature goes on all over the planet, from Brazil to Zimbabwe, but we still have not explained fully why people care or are moved to act. One set of explanations derives from examples like Ashoka, the ancient ruler of India (third century B.C.), who set aside the world's first wildlife preserve after converting to Buddhism and its doctrine of *ahimsa*, or nonviolence toward all living things. Ashoka's case suggests either that a traditional religion like Buddhism has been the driving force or that powerful elites deserve credit for protecting the natural world. Both explanations can claim a degree of truth. But the most active nations in nature protection have not been especially devoted to Buddhism or to other traditional faiths, while most elites, from emperors to corporate executives, have been destructive of or indifferent to nature.

Elites may, of course, be moved by a love of nature as much as are non-elites, just as they may show a genuine concern for their less affluent fellow citizens and for the welfare of future generations. That people of wealth and high status have played a role in preserving nature cannot be denied, although often their efforts have aimed to secure good hunting or exotic travel for themselves and have not been devoted to a more altruistic love of nature. The preservation of African wildlife by Europeans, for example, grew out of motives of class privilege mixed with a more egalitarian concern for human well-being and the needs of other species.[1]

Ordinary individuals, on the other hand, have thrilled as much as elites to the smell of a forest or the sight of a wild antelope and have been moved as much by scenes of natural beauty. The protection of nature owes a great deal to them too. In fact, I will argue that they have been far more important than historians have commonly acknowledged in bringing change in environmental attitudes, including attitudes toward wilderness. We have not fully appreciated how much the protection of wild nature owes to the spread of modern liberal, democratic ideals and to the support of millions of ordinary people around the world.

The role of democracy in promoting nature protection becomes clear when we examine where most of that protection has occurred in the modern world. Overwhelmingly, it has taken place within nations that profess democratic principles, cherish human rights, and allow freedom of speech and dissent from official dogma. Wherever open, egalitarian societies have taken root, protection has spread rapidly. Conversely, it has generally failed when confronted by powerful technocrats, politburos, and other religious or political forms of authoritarianism.

Fortunately for a world undergoing a continuing democratic revolution, there is plenty of wild nature left to protect. In 1989, a reconnaissance survey found that 48 million square kilometers of the planet qualified as wilderness, or about a third of the total land surface.[2] (Forty-eight million square kilometers is equivalent to 12 billion acres, an expanse larger than the Western Hemisphere.) Fifteen million of those square kilometers are in Antarctica and Greenland—vast white wildernesses of ice.[3] Much of the earth's surface in the higher latitudes is remarkably wild, as are much of the world's deserts and tropical rainforests and virtually all of the oceans, where until very recently there have been few traces of human impact.

Traditionalists might insist that wilderness must mean forested mountains, not glaciers or oceans, but that would be a highly arbitrary definition. Wilderness as defined by the survey does not refer to a particular kind of biome; it can include any sort of nature, whether forest, grassland, desert, polar ice cap, volcanic plain, lake, or sea, that shows little sign of active human settlement or commodity production.

The 1989 survey looked for areas larger than 400,000 hectares (1 million acres) that lacked any "permanent human settlements or roads," lands that were "not regularly cultivated nor heavily and continuously grazed" but that might have been "lightly used and occupied by indigenous peoples at various times who practiced traditional subsistence styles of life." Wilderness purists might not like the looseness of that standard; for some, a single tissue can spoil a place, or a solitary fisherman's hook, or a lone donkey track. For those *opposed* to any strict protection, on the other hand, even the light passage of a primitive tribe through the landscape should disqualify it as wilderness, and they want to put it in the category of a "well-used" or "managed" place, open to exploitation. Neither kind of absolutism will do; neither reflects the flexible, pragmatic definitions people have historically used or the inescapable relativity of the term "wilderness."[4]

The United States, despite its persistent frontier image, ranked low on the list of wilderness-rich countries—down at number sixteen, with only 440,580 square kilometers (109 million acres), or 4.7 percent of its total area. Higher on the list were Russia, Canada, Australia, Brazil, the Sudan, and Algeria. Several heavily populated countries were surprisingly high on the list, including China, India, Laos, Mexico, and Iraq. China, for instance, despite its 1 billion-plus population, still had 22 percent of its territory in a wild state, a far higher percentage than the United States.

A survey that focuses only on huge, million-acre parcels of land does not, of course, exhaust the possibilities of wild places on the earth. There are many places under that size that might qualify as wild—a mere ten thousand or a hundred thousand acres in extent. And then there are all those smaller, even tiny, patches of wildness that lurk on the edges of our cities, farms, and backyards and that may be wonderfully rich in diversity and high in aesthetic and spiritual value.

Where the United States stands high among nations, where it might even be called exceptional, is not so much in the extent of its remaining wildlands as in its long history of activism in protecting them. The United States was the first nation to create a national park (in 1864 or 1872, depending on whether one grants priority to Yosemite or Yellowstone), the first to set up a full-blown wilderness preservation system (1964), and the first to pass an endangered species act (1973).[5]

That historic leadership role seems to have come to a fitful end, following the defeat in 1980 of President Jimmy Carter, who managed in his last month in office to sign protection for more than 100 million acres of Alaska's wildlands, and then following the departure of the Clinton presidency, which in one magnificent moment declared an additional 58.5 million acres of U.S. forest lands to be forever free of roads. That Clinton ruling was quickly suspended by the second Bush administration. In recent years, the cause of wildlands protection has been rejected on the political Right for stifling private enterprise and has been criticized by some on the political Left for detracting attention from issues of social justice.[6] As a consequence, leadership in nature protection has passed to other nations, some of which are the older democracies while others are relatively younger nations still struggling to transfer more power to the people and to make nature preservation part of their culture.

That shift in leadership was noticeable at the 1992 Earth Summit in Rio de Janeiro when the United States took a back seat as more than a hundred nations agreed that every country should protect at least 12 percent of its land base from economic use. Not every nation voting at that meeting was a full-fledged democracy, but the decision was one that reflected a democratic process of open discussion and global representation. It was animated by an egalitarian purpose—to protect the beauty, health, and integrity of nature for the sake of future human generations and to recognize a moral obligation to save other forms of life from extinction.

Preservationists all over the world have agreed on a common program to set up protective zones where farming, logging, mining, town building, wildlife poaching, or the dumping of wastes is prohibited or severely restricted. They represent a wide array of ethnic backgrounds and languages. The Nordic countries, for example, have produced plenty of activists and can boast some of the most carefully protected wildlands in the world. Thousands of miles away, and sharply contrasting in many ways, is Costa Rica, which has protected 28 percent of its territory from development—11 percent in national parks, 4 percent in indigenous reserves, and 13 percent in a miscellaneous series of biological reserves, national forests, national monuments, and national wildlife refuges.[7] The spectacular diversity of its flora and fauna, the stunning beauty of its mountain ranges, exuberant wet and dry forests, and broad saltwater beaches have given rise to one of the world's most conservation-minded societies. Next door, Panama in its post-Noriega period is moving toward a similar policy of large-scale, vigorous nature protection. What joins those two Central American countries to Norway, Finland, or Sweden, or joins any of them to New Zealand and its great protected wilderness of Milford Sound? Why are many other nations so backward in preserving wild places—Russia, for example, or Guatemala or Thailand?

The conventional answer is that preserving nature appeals only to affluent people whose stomachs are full and is never important to the poor or the aspiring. At the extremes, this seems to be true; desperately hungry men and women are not likely to think much about wilderness or, indeed, think much about many other large issues at the national or global scales. But such an economic explanation is too simplistic and reductive to be dependable. Income alone does not work very well *within* societies in predicting which citizens care about preservation and which do not; it cannot explain why some oil executives care while plenty of others do not, nor, on the other hand, why some pensioners care while others do not.

Nor does a simple economic explanation work at the international level. According to World Bank data from 2003, Norway stands third in the world in gross national income per capita (US$43,400); Finland, thirteenth ($27,060); New Zealand, fortieth ($15,530); Costa Rica, seventy-seventh ($4,300); and Panama, seventy-ninth ($4,060).[8] Huge differences in wealth, yet all are active countries in nature awareness and preservation. Furthermore, within the most abysmally poor countries, where there may be little or no organized movement for preservation, many people care deeply about wildlife and unspoiled natural beauty.

More reliable indicators of whether nations become active in preserving wild places are the state of personal freedom, the degree of social equality, and the sanctity of human rights. Far more than religious or ethnic identity or gross national product, the quality of nature protection seems to correlate with the quality of democracy. Countries where there is a more equitable distribution of economic opportunity, a low level of militarism, high levels of literacy, greater racial and gender equality, free and competitive elections, and tolerance of dissent tend to set aside significant pieces of nature for protection from economic development. Or, if they are too densely settled for that to be a realistic possibility at home, they work to do the same internationally—as Denmark has recently done in setting aside much of Greenland as the world's largest national park. Why that should be so, why liberal democracy should correlate to wildlands protection, is a question that has never been fully explored, although it is of the utmost importance to the future of life on earth.

The history and meaning of liberal democracy is an old and complicated subject. We have come to realize that it refers to more than the superficial mechanics of political modernization—elections, parliaments, or governmental checks and balances. Liberal democracy is founded on a pair of intertwined cultural ideals: personal liberty and social equality. The greatest proponent of those ideals was Jean-Jacques Rousseau (1712–1778), who insisted that one ideal could not exist without the other. A government that promises equality to its citizens will never deliver that condition without the constant pressure of free, critical, and dissenting opinion. Equality needs liberty, and liberty needs equality. That linkage has often come under challenge by those who want to promote one ideal but not the other: for example, political philosophers like Alexis de Tocqueville, author of *Democracy in America* (1835–1840), who preferred liberty over equality, or politicians like Mao Zedong (communist dictator of China from 1949 to 1976), who sacrificed liberty in pursuit of the classless state. But the critics have not succeeded in splitting them apart. The two ideals have not always been easy to reconcile, but together they have worked to change the course of Western history and, increasingly, to change the dynamics of non-Western societies as well.

Much has been written on how that pair of ideals has revolutionized human relations but rather less on how they have affected people's relation to nature.[9] Their environmental impact has been little short of revolutionary too. Old notions that humans have been created specially in the image of God or that they have been given dominion over all other forms of life or that they can draw a rigid line around their own liberty or equality, making those ideals exclusive to *Homo sapiens*, have proved unsustainable. Nature has become the patron and partner of liberal democracy. It has even come to be seen as the source of human liberation, a place of freedom and of equality, and therefore worthy of respect, protection, and even worship.

"I wish to speak a word for Nature," declared Henry David Thoreau in 1862, "for absolute freedom and wildness, as contrasted with a freedom and culture merely civil."[10] Going into wild country, as Thoreau advocated, experiencing places free of human domination, becomes a means of freeing oneself from the hand of convention or authority. Social deference fades in the wilderness. Economic rank does not matter so much. Money is not needed to survive there. Nature offers a home to the dissident mind, the rebellious child, the outlaw, the runaway slave, the soldier who refuses to fight, and (by the late nineteenth century) the woman who goes mountain climbing to show her strength and independence.[11]

A move toward greater equality among species became irresistible too, giving rise to animal rights, wildlife refuges, and even Darwin's theory of evolution, which joined humans and other forms of life into a common brotherhood. Plants and animals came to be valued for more than their potential for domestication, their fitness for pulling a wagon or yielding a crop; wild species came to be admired for surviving on their own, independent of human purposes. They were seen to form their own communities. They were not inferior versions of ourselves, but beings created by God or evolving by natural processes for their own sakes. They were, as John Muir argued (echoing the Scottish poet Robert Burns), "earth-born companions and our fellow mortals."[12]

Nature in the wake of liberal democracy also became the basis of a new (or rediscovered) religion, a fathomless source of spirituality, complementary to or independent of traditional religion. Woods, mountains, or prairies became divine texts in which one could find answers to life's ultimate questions, without the mediation of church authorities or theologians. Protestants in Western Europe led the way to this new religion by challenging the entrenched hegemony of the pope and the Roman Catholic church and by insisting that every individual has a right and duty to read the Holy Bible for her- or himself. They, and particularly groups like the Quakers and Presbyterians, opened a challenge to hierarchical religion that they then had trouble controlling within their own denominational walls. Any written Bible or testament came to be seen as a manmade artifact full of human frailties and limitations, inferior to the outdoors as a source of inspiration. Nature drew people away from all established creeds and faiths. In the presence of nature, the rising liberal spirit of the nineteenth and twentieth centuries found a new source of guidance accessible to any individual.

One of the great pioneers of that new religion of nature was Rousseau. In 1762, both the French and Swiss governments threatened him with arrest for being a dangerous heretic, a radical, and anti-Christian. Seeking refuge from the authorities on St. Pierre's Island in the Bieler See near Berne, Switzerland, he immersed himself, body and mind, in the wholeness of nature. His memoir, *The Reveries of the Solitary Walker*, tells about finding a subversive source of spiritual insight:

> The earth, in the harmony of the three realms [mineral, plant, animal], offers man a spectacle filled with life, interest, and charm—the only spectacle in the world of which his eyes and his heart never weary. The more sensitive a soul a contemplator has, the more he gives himself up to the ecstasies this harmony arouses in him. A sweet and deep reverie takes

> possession of his senses then, and through a delightful intoxication he loses himself in the immensity of this beautiful system with which he feels himself one.[13]

Others have felt this call of nature too, from William Wordsworth and Johann Goethe down to Rachel Carson and Robert Marshall. Whatever their national or religious roots, they have broken free from orthodoxy and found in nature part or all of what they needed to feed their spiritual hunger.

If the nature protection movement has arisen in the wake of liberal democracy, influenced deeply by the ideals of liberty and equality, then we should not expect to find that movement blooming in places where repressive authority and inequality stand in the way. We should not expect a preservation ethic to flourish in men like Anastasio Somoza, the dictatorial president of Nicaragua during the 1940s and 1950s, or Colonel Joseph-Désiré Mobutu, the kleptocratic strong man of Zaire until he was deposed in 1997, or in such totalitarians as Mao, Stalin, or Pol Pot. We should not be surprised that wildlands are not attractive to military juntas, theocracies, patriarchies, or slave regimes. Nature in its wilder state is a threat to such authoritarian minds. It is where danger lurks, threatening always to erupt and bring down their vulnerable edifices of control.

We should expect, on the other hand, that nations in the forefront of nature preservation would be those influenced by ideals of liberty and equality, and indeed that is so: Costa Rica and Panama in Central America; New Zealand and Australia in the South Pacific; the United States and Canada in North America; Norway, Scotland, and others in Europe—and, eventually, a new Bulgaria, Chile, India, Zambia, or other nation where the quality of democracy shows significant advance.

Liberal democracies are, of course, more than expressions of cultural ideals. They are also systems of governance, and to do that work they must pass laws and regulations. In doing so, they must infringe on the liberty of some citizens in order to protect the liberties of others, or to protect the spiritual values of wilderness or the rights of other species to survive. This rule making can lead to charges of injustice, and sometimes the charges are justified. Liberal democracies, in their making of laws and regulations, have not been free of class, gender, or racial bias or always respectful of differences of opinion. They are imperfect creations. Tocqueville rightly warned about some of their shortcomings: a tendency toward tyranny of the majority exercised over minorities, a tendency to glorify greed (under the doctrine of economic liberalism), and a susceptibility to elites who gain power through the free and ruthless accumulation of money. Those who demand their own freedom can be quick to deny it to others. Despite

Rousseau's confidence that virtue must always flourish where liberty and equality together flourish, history shows a more complicated picture: liberal democracies that display hypocrisy along with virtue, conflict as well as cooperation, and bigotry of all sorts.

Similarly, the record of liberal democracies in protecting nature has often been flawed by narrow self-interest. The pursuit of liberty has at times meant the freedom to invade and exploit the natural world for personal gain. The pursuit of equality, for all of its positive appeal, has often led to environmental destruction; it has been one of the driving forces behind modern consumer culture, which promises everyone a more abundant material life and endless economic growth, regardless of the ecological consequences. Here again are contradictions difficult to resolve and impossible to avoid. Those contradictions have driven much of modern history. It is precisely because of them—the tensions between liberty and equality, between present and future generations' claims on the earth, and between human rights and nature's right to exist—that liberal democracies do not represent some ultimate or unstoppable victory. They are not, at least in their current forms, the "end of history."[14]

The most serious challenges facing conservationists are those regimes that have never been touched by or are falling back from liberal democratic ideals. They are many, and they control the destiny of much of the remaining wilderness on the planet. Some are still locked in repressive structures of power that allow no dissent from orthodoxy, no openness to new ideas or research, and no respect for the other-than-human world. Then there is the challenge of nations where liberal democracy is weakening or failing, as authoritarian forces within them gain strength. They too are not hard to find: look for imperial-scale military budgets, social intolerance, education giving way to indoctrination, oil drilling in the last wild places, and dark warnings against "pagan" heresies. Look not only to sinister foreign places; look within the United States and other countries where there are growing internal forces of reaction and repression.

Perhaps this is the way that the dream of liberal democracy self-destructs: in their quest for freedom and equality, people may devour the earth before they save it, and in devouring the earth they may lose the freedom they thought they were getting. They may end up as slaves to their own appetites, living in fearful bondage to whatever ideas or forces will offer them security. That seems to describe accurately the current mood of many Americans and others around the world.

But the historic association of nature protection with the spread of liberty and equality has been a strong force, reaching into almost every corner of this imperiled planet. It may prove to be more powerful than any backlash, and it may push all nations toward a global ethic of conservation.

Notes

1. These conflicts are well discussed in Jane Carruthers's work on South Africa, *The Kruger National Park: A Social and Political History* (Pietermaritzburg, South Africa: Natal University Press, 1995).

2. J. Michael McCloskey and Heather Spalding, "A Reconnaissance-Level Inventory of the Amount of Wilderness Remaining in the World," *Ambio* 18 (1989): 221–27. The survey did not include the 70 percent of the earth's surface covered by oceans, most of which is hardly explored in depth, let alone domesticated.

3. According to the United Nations Environment Program, approximately 19 million square kilometers, an area the size of Canada and the United States combined, have been given some protection globally. Eleven percent of that total, or 2 million square kilometers, has been placed under "strict" protection as a wilderness or nature reserve. See Table 1, "2003 United Nations List of Protected Areas," at http://www.unep-wcmc.org.

4. A more difficult challenge for any definition of wilderness comes from the new potential for anthropogenic change in the global climate system. But even if we grant such change as scientific fact, it does not follow that we should now call every place on earth a "cultural landscape." A cultural landscape has been deliberately shaped by ideas and values, while global warming, anthropogenic or not, is as unwitting and unpredictable as a meteor hitting the earth. Moreover, a land left free of ice by global warming may still be "wild" if it is unsettled by human population or unexploited for commodities.

5. In some places, protection commenced almost as long ago as in the United States: in 1894, for example, the Maori leader Te Heuheu Tukino IV gave the austere volcanic peak region of the North Island, the Tongariro National Park, to the nation of New Zealand, and ten years later that country made the fjord lands of the South Island off-limits to economic development. See Paul Star and Lynne Lochhead, "Children of the Burnt Bush: New Zealanders and the Indigenous Remnant, 1880–1930," in *Environmental Histories of New Zealand*, ed. Eric Pawson and Tom Brooking (Melbourne: Oxford University Press, 2002), 123–27.

6. For a critique of this debunking spirit among historians see my essay "The Wilderness of History," *Wild Earth* 7 (Fall 1997): 9–13.

7. Sterling Evans, *The Green Republic: A Conservation History of Costa Rica* (Austin: University of Texas Press, 1999), 7–8.

8. See the comparative tables on per capita gross national income (GNI) at the World Bank Web site: http://www.worldbank.org/data/quickreference/quickref.html.

9. An exception to this observation is Roderick Nash, *The Rights of Nature: A History of Environmental Ethics* (Madison: University of Wisconsin Press, 1989). Although focused mainly on liberal democratic ideals within the United States, Nash does include such figures as the Norwegian Arne Naess, founder of the deep ecology movement, whose ideas seem profoundly indebted to Rousseau, William Wordsworth, and other early modern thinkers.

10. Thoreau, "Walking," *Atlantic Monthly* 9 (June 1862): 657.

11. See, for example, Susan R. Schrepfer, *Nature's Altars: Mountains, Gender, and American Environmentalism* (Lawrence: University Press of Kansas, 2005).

12. Muir, *A Thousand Mile Walk to the Gulf* (Boston: Houghton Mifflin, 1916), 139.

13. Rousseau, *The Reveries of the Solitary Walker*, in *The Collected Writings of Jean-Jacques Rousseau*, vol. 8, ed. Christopher Kelly and trans. Charles Butterworth (Hanover, N.H.: University Press of New England, 2000), 59.

14. The environmental movement has laid bare those tensions within liberal democracy, although surprisingly it gets little credit for doing so in Francis Fukuyama's *The End of History and the Last Man* (New York: Free Press, 1992).

Recommended Readings

Each chapter includes its own notes. This is not a comprehensive bibliography for this volume, but rather a selection of key primary and secondary works.

Abbey, Edward. *Desert Solitaire: A Season in the Wilderness.* New York: McGraw-Hill, 1968.

———. *The Journey Home: Some Words in Defense of the American West.* New York: Penguin, 1977.

Adams, William M., and Martin Mulligan, eds. *Decolonizing Nature: Strategies for Conservation in a Post-Colonial Era.* London: Earthscan, 2003.

Anderson, Virginia DeJohn. *Creatures of Empire: How Domestic Animals Transformed Early America.* New York: Oxford University Press, 2004.

Backes, David. *A Wilderness Within: The Life of Sigurd F. Olson.* Minneapolis: University of Minnesota Press.

Balée, William, ed. *Advances in Historical Ecology.* New York: Columbia University Press, 1998.

Bederman, Gail. *Manliness and Civilization: A Cultural History of Gender and Race in the United States, 1880–1917.* Chicago: University of Chicago Press, 1995.

Beinart, William, and Peter Coates. *Environment and History: The Taming of Nature in the USA and South Africa.* New York: Routledge, 1995.

Brockington, Dan. *Fortress Conservation: The Preservation of the Mkomazi Game Reserve.* Oxford: Currey, 2002.

Brody, Hugh. *The Other Side of Eden: Hunters, Farmers, and the Shaping of the World.* Vancouver, B.C.: Douglas & McIntyre, 2000.

Brower, David. *Work in Progress.* Salt Lake City, Utah: Peregrine Smith, 1991.

Bryson, Michael. *Visions of the Land: Science, Literature, and the American Environment from the Era of Exploration to the Age of Ecology.* Charlottesville: University Press of Virginia, 2002.

Budiansky, Stephen. *Nature's Keepers: The New Science of Nature Management.* New York: Free Press, 1995.

Burnham, Philip. *Indian Country, God's Country: Native Americans and the National Parks.* Washington, D.C.: Island, 2000.

Callicott, J. Baird, and Susan Flader, eds. *The River of the Mother of God and Other Essays by Aldo Leopold.* Madison: University of Wisconsin Press, 1991.

Callicott, J. Baird, and Michael Nelson, eds. *The Great New Wilderness Debate.* Athens: University of Georgia Press, 1998.

Canup, John. *Out of the Wilderness.* Middletown, Conn.: Wesleyan University Press, 1990.

Carruthers, Jane. *The Kruger National Park: A Social and Political History.* Pietermaritzburg, South Africa: Natal University Press, 1995.

Carson, Rachel. *Silent Spring.* Boston: Houghton Mifflin, 1962.

———. *The Sea: The Sea around Us; Under the Sea-Wind; The Edge of the Sea.* London: Paladin, 1968.

Catton, Theodore. *Inhabited Wilderness: Indians, Eskimos, and the National Parks in Alaska.* Albuquerque: University of New Mexico Press, 1997.

Cawley, R. McGreggor. *Federal Land, Western Anger: The Sagebrush Rebellion and Environmental Politics.* Lawrence: University Press of Kansas, 1993.

Cohen, Michael P. *The Pathless Way: John Muir and American Wilderness.* Madison: University of Wisconsin Press, 1984.

———. *The History of the Sierra Club, 1892–1970.* San Francisco, Calif.: Sierra Club Books, 1988.

Commoner, Barry. *The Closing Circle: Nature, Man, and Technology.* New York: Knopf, 1971.

Conte, Christopher. *Highland Sanctuary: Environment and History in Tanzania's Usambara Mountains.* Athens: Ohio University Press, 2004.

Cooper, Susan. *Rural Hours.* Boston: Houghton Mifflin, 1887.

Croker, Robert. *Pioneer Ecologist: The Life and Work of Victor Ernest Shelford, 1877–1968.* Washington, D.C.: Smithsonian Institution Press, 1991.

Cronon, William. *Changes in the Land.* New York: Hill and Wang, 1983.

———. *Nature's Metropolis: Chicago and the Great West.* New York: Norton, 1991.

Cronon, William, ed. *Uncommon Ground: Toward Reinventing Nature.* New York: Norton, 1995.

Cutright, Paul Russell. *Theodore Roosevelt: The Making of a Conservationist.* Urbana: University of Illinois Press, 1985.

DeBuys, William. *Enchantment and Exploitation: The Life and Hard Times of a New Mexico Mountain Range.* Albuquerque: University of New Mexico Press, 1985.

Donahue, Brian. *The Great Meadow: Farmers and the Land in Colonial Concord.* New Haven, Conn.: Yale University Press, 2004.

Dunaway, Finis. *Natural Visions: The Power of Images in American Environmental Reform.* Chicago: University of Chicago Press, 2005.

Evans, Sterling. *The Green Republic: A Conservation History of Costa Rica.* Austin: University of Texas Press, 1999.

Fagen, Brian. *The Little Ice Age: How Climate Made History.* New York: Basic, 2001.

Fairhead, James, and Melissa Leach. *Misreading the African Landscape: Society and Ecology in a Forest-Savanna Mosaic.* Cambridge: Cambridge University Press, 1996.

Flader, Susan. *Thinking Like a Mountain: Aldo Leopold and the Evolution of an Ecological Attitude toward Deer, Wolves, and Forests.* Columbia: University of Missouri Press, 1974.

Flippen, J. Brooks. *Nixon and the Environment.* Albuquerque: University of New Mexico Press, 2000.

Foreman, Dave. *Confessions of an Eco-Warrior.* New York: Harmony, 1991.

———. *Rewilding North America: A Vision for Conservation in the 21st Century.* Washington, D.C.: Island, 2004.

Fox, Stephen R. *The American Conservation Movement.* Madison: University of Wisconsin Press, 1986.

Frome, Michael. *Battle for the Wilderness,* rev. ed. Salt Lake City: University of Utah Press, 1997.

Gottlieb, Robert. *Forcing the Spring: The Transformation of the American Environmental Movement.* Washington, D.C.: Island, 1993.

Graf, William L. *Wilderness Preservation and the Sagebrush Rebellions.* Savage, Md.: Rowman & Littlefield, 1990.

Greenough, Paul, and Anna Lowenhaupt Tsing, eds. *Nature in the Global South: Environmental Projects in South and Southeast Asia.* Durham, N.C.: Duke University Press, 2003.

Griffiths, Tom, and Libby Robin, eds. *Ecology and Empire: Environmental History of Settler Societies.* Seattle: University of Washington Press, 1997.

Guha, Ramachandra. *Environmentalism: A Global History.* New York: Addison, Wesley Longman, 2000.

Gutiérrez, Ramón A. *When Jesus Came, the Corn Mothers Went Away: Marriage, Sexuality, and Power in New Mexico, 1500–1846.* Stanford, Calif.: Stanford University Press, 1991.

Harrison, Robert Pogue. *Forests: The Shadow of Civilization.* Chicago: University of Chicago Press, 1992.

Harvey, Mark W. T. *A Symbol of Wilderness: Echo Park and the American Conservation Movement.* Albuquerque: University of New Mexico Press, 1994; rpt., Seattle: University of Washington Press, 2000.

———. *Wilderness Forever: Howard Zahniser and the Path to the Wilderness Act.* Seattle: University of Washington Press, 2006.

Hayes, Samuel P. *Conservation and the Gospel of Efficiency: The Progressive Conservation Movement, 1890–1920.* Cambridge, Mass.: Harvard University Press, 1959.

Hill, Julia Butterfly. *The Legacy of Luna: The Story of a Tree, a Woman and the Struggle to Save the Redwoods.* San Francisco, Calif.: Harper San Francisco, 2000.

Hirt, Paul. *A Conspiracy of Optimism: Management of the National Forests since World War Two.* Lincoln: University of Nebraska Press, 1994.

Holmes, Steven J. *The Young John Muir: An Environmental Biography.* Madison: University of Wisconsin Press, 1999.

Isenberg, Andrew C. *The Destruction of the Bison: An Environmental History, 1750–1920.* New York: Cambridge University Press, 2000.

Jacoby, Karl. *Crimes against Nature: Squatters, Poachers, Thieves, and the Hidden History of American Conservation.* Berkeley: University of California Press, 2001.

Jarvis, Kimberly A. *Nature and Identity in the Creation of Franconia Notch.* Hanover: University Press of New England, 2007.

Johnson, Douglas, and David Anderson, eds. *The Ecology of Survival.* Boulder, Colo.: Westview, 1988.

Keller, Robert H., and Michael F. Turek. *American Indians and National Parks.* Tucson: University of Arizona Press, 1998.

Kennedy, Roger G. *Mr. Jefferson's Lost Cause: Land, Farmers, Slavery, and the Louisiana Purchase.* New York: Oxford University Press, 2003.

Kinsey, Joni Louise. *Thomas Moran and the Surveying of the American West.* Washington, D.C.: Smithsonian Institution Press, 1992.

Knobloch, Frieda. *The Culture of Wilderness: Agriculture as Colonization in the American West.* Chapel Hill: University of North Carolina Press, 1996.

Krech, Shepherd, III. *The Ecological Indian.* New York: Norton, 1999.

Lamar, Howard, and Leonard Thompson, eds. *The Frontier in History: North America and Southern Africa Compared.* New Haven, Conn.: Yale University Press, 1981.

Lane, Belden. *Landscapes of the Sacred,* 2d ed. Baltimore, Md.: Johns Hopkins University Press, 2001.

Leopold, Aldo. *A Sand County Almanac.* New York: Oxford University Press, 1948.

Lewis, Michael. *Inventing Global Ecology: Tracking the Biodiversity Ideal in India, 1947–1997.* Athens: Ohio University Press, 2004.

Logan, Michael. *The Lessening Stream.* Tucson: University of Arizona Press, 2002.

Louter, David. *Windshield Wilderness: Cars, Roads, and Nature in Washington's National Parks.* Seattle: University of Washington Press, 2006.

Lowenthal, David. *George Perkins Marsh: Prophet of Conservation.* Seattle: University of Washington Press, 2000.

MacKenzie, John. *The Empire of Nature: Hunting, Conservation and British Imperialism.* Manchester, England: University of Manchester Press, 1988.

Marsh, George Perkins. *Man and Nature.* Cambridge, Mass.: Harvard University Press, 1965.

McGregor, Robert Kuhn. *A Wider View of the Universe: Henry Thoreau's Study of Nature.* Urbana: University of Illinois Press, 1997.

McKibben, Bill. *The End of Nature.* New York: Anchor, 1997.

McNeill, John R. *Something New under the Sun: An Environmental History of the Twentieth Century World.* New York: Norton, 2000.

Meine, Curt. *Aldo Leopold: His Life and Work.* Madison: University of Wisconsin Press, 1988.

Melville, Elinor G. K. *A Plague of Sheep: Environmental Consequences of the Conquest of Mexico.* New York: Cambridge University Press, 1994.

Merchant, Carolyn. *Ecological Revolutions: Nature, Gender, and Science in New England.* Chapel Hill: University of North Carolina Press, 1989.

———. *Earthcare: Women and the Environment.* New York: Routledge, 1996.

———. *The Columbia Guide to American Environmental History.* New York: Columbia University Press, 2002.

———. *Reinventing Eden: The Fate of Nature in Western Culture.* New York: Routledge, 2004.

Miles, John. *Guardians of the Parks: A History of the National Parks and Conservation Association.* Washington, D.C.: Taylor & Francis, 1995.

Miller, Angela. *Empire of the Eye: Landscape Representation and American Cultural Politics, 1825–1875.* Ithaca, N.Y.: Cornell University Press, 1996.

Miller, Char. *Gifford Pinchot and the Making of Modern Environmentalism.* Washington, D.C.: Shearwater, 2001.

Miller, Char, ed. *Fluid Arguments: Five Centuries of Western Water Conflict.* Tucson: University of Arizona Press, 2001.

Miller, Sally M., ed. *John Muir: Life and Work.* Albuquerque: University of New Mexico Press, 1993.

Mitman, Gregg. *Reel Nature: America's Romance with Wildlife on Film.* Cambridge, Mass.: Harvard University Press, 1999.

Muir, John. *John Muir: Nature Writings.* New York: Literary Classics of America, 1997.

Nabhan, Gary. *Cultures of Habitat: On Nature, Culture and Story.* Tucson, Ariz.: Treasure Chest, 1998.

———. *The Desert Smells Like Rain: A Naturalist in O'Odham Country.* Tucson: University of Arizona Press, 2002.

Nash, Roderick. *The Rights of Nature: A History of Environmental Ethics.* Madison: University of Wisconsin Press, 1989.

———. *Wilderness and the American Mind,* 4th ed. New Haven, Conn.: Yale University Press, 2001.

Norwood, Vera. *Made from This Earth: American Women and Nature.* Chapel Hill: University of North Carolina Press, 1993.

Olson, Sigurd. *The Singing Wilderness.* New York: Knopf, 1956.

Pawson, Eric, and Tom Brooking, eds. *Environmental Histories of New Zealand.* Melbourne, Australia: Oxford University Press, 2002.

Pearson, Byron. *Still the River Runs: Congress, the Sierra Club, and the Fight to Save Grand Canyon.* Tucson: University of Arizona Press, 2002.

Pollan, Michael. *Second Nature: A Gardener's Education.* New York: Dell, 1991.

Price, Jennifer. *Flight Maps: Adventures with Nature in Modern America.* New York: Basic, 1999.

Pyne, Stephen. *Fire in America: A Cultural History of Woodland and Rural Fire.* Princeton, N.J.: Princeton University Press, 1982.

Richards, John F. *The Unending Frontier: An Environmental History of the Early Modern World.* Berkeley and Los Angeles: University of California Press, 2003.

Richter, Daniel. *Facing East from Indian Country.* Cambridge, Mass.: Harvard University Press, 2001.

Rome, Adam. *The Bulldozer in the Countryside: Suburban Sprawl and the Rise of American Environmentalism.* Cambridge: Cambridge University Press, 2001.

Roth, Dennis M. *The Wilderness Movement and the National Forests: 1964–1980.* Washington, D.C.: GPO, 1984.

Rothman, Hal K. *The Greening of a Nation? Environmentalism in the United States since 1945.* New York: Harcourt Brace, 1998.

Scharff, Virginia. *Seeing Nature through Gender*. Lawrence: University Press of Kansas, 2003.

Schrepfer, Susan R. *Nature's Altars: Mountains, Gender, and American Environmentalism.* Lawrence: University Press of Kansas, 2005.

Scott, Doug. *The Enduring Wilderness.* Golden, Colo.: Fulcrum, 2004.

Shaffer, Marguerite. *See America First: Tourism and National Identity, 1880–1940.* Washington, D.C.: Smithsonian Institution Press, 2001.

Slaughter, Thomas P. *The Natures of John and William Bartram.* New York: Knopf, 1997.

Smith, Thomas G. *Green Republican: John Saylor and the Preservation of America's Wilderness.* Pittsburgh, Pa.: University of Pittsburgh Press, 2006.

Solnit, Rebecca. *Savage Dreams: A Journey into the Landscape Wars of the American West.* Berkeley: University of California Press, 1999.

Soulé, Michael, and B. A. Wilcox. *Conservation Biology: An Evolutionary-Ecological Perspective.* Sunderland, Mass.: Sinauer, 1980.

Soulé, Michael, and Gary Lease, eds. *Reinventing Nature? Responses to Postmodern Deconstruction.* Washington, D.C.: Island, 1995.

Spaulding, Jonathan. *Ansel Adams and the American Landscape.* Berkeley and Los Angeles: University of California Press, 1995.

Spence, Mark David. *Dispossessing the Wilderness: Indian Removal and the Making of the National Parks.* New York: Oxford University Press, 1999.

Stegner, Wallace, ed. *This Is Dinosaur: Echo Park Country and Its Magic Rivers.* New York: Knopf, 1955.

Steinberg, Ted. *Down to Earth: Nature's Role in American History.* New York: Oxford University Press, 2002.

Stevens, Stan, ed. *Conservation through Cultural Survival: Indigenous People and Protected Areas.* Washington, D.C.: Island, 1997.

Stewart, Omer. *Forgotten Fires.* Norman: University of Oklahoma Press, 2002.

Stoll, Mark. *Protestantism, Capitalism, and Nature in America.* Albuquerque: University of New Mexico Press, 1997.

Stoll, Steven. *Larding the Lean Earth: Soil and Society in Nineteenth-Century America.* New York: Hill and Wang, 2002.

Sutter, Paul. *Driven Wild: How the Fight against Automobiles Launched the Modern Wilderness Movement.* Seattle: University of Washington Press, 2002.

Switzer, Jacqueline. *Green Backlash: The History and Politics of Environmental Opposition in the United States.* Boulder, Colo.: Rienner, 1997.

Takacs, David. *The Idea of Biodiversity.* Baltimore, Md.: John Hopkins University Press, 1996.

Thoreau, Henry D. *Faith in a Seed*, ed. Bradley P. Dean. Covelo, Calif.: Shearwater Books/Island Press, 1993.

———. *Letters to a Spiritual Seeker*, ed. Bradley P. Dean. New York: Norton, 2004.

Turner, Frederick Jackson. *The Frontier in American History.* New York: Holt, Rinehart and Winston, 1920.

Turner, James Morton. *The Promise of Wilderness: A History of American Environmental Politics.* Seattle: University of Washington Press. In press.

Warren, Louis. *The Hunter's Game: Poachers and Conservationists in Twentieth-Century America.* New Haven, Conn.: Yale University Press, 1997.

Wayburn, Edgar, with Allison Alsup. *Your Land and Mine: Evolution of a Conservationist.* San Francisco, Calif.: Sierra Club Books, 2004.

Whitmore, Thomas, and B. L. Turner. *Cultivated Landscapes of Middle America on the Eve of Conquest.* Oxford: Oxford University Press, 2001.

Wilson, E. O. *Naturalist.* Washington, D.C.: Island, 1994.

Worster, Donald. *Nature's Economy: A History of Ecological Ideas*, 2d ed. New York: Cambridge University Press, 1994.

———. *A River Running West: The Life of John Wesley Powell.* New York: Oxford University Press, 2001.

Worster, Donald, ed. *The Ends of the Earth: Perspectives on Modern Environmental History.* Cambridge: Cambridge University Press, 1988.

Zakin, Susan. *Coyotes and Town Dogs: Earth First! and the Environmental Movement.* New York: Viking, 1993.

Index